A nostalgic look at

GLASGOW
TRAMS

Since 1950

A classic Glasgow scene with two classic Glasgow 'Standard' tramcars. Both are pictured near Glasgow Cross in April 1959 outside the Brighter Homes shop on Trongate. No. 479 is, by this time, something of a veteran, having been built in 1903, whereas No. 49 is a positive youngster, having first seen the light of day in 1921! Both were subsequently withdrawn from service in that April, their longevity, together with that of their sister cars, being a fine testament to their builders. Service 15 was at one time one of the longest trips on the system, running from Airdrie to Elderslie. It was subsequently cut back over the years until finally its journey was from Anderston Cross to Shettleston or Baillieston. *(Bob Mack Collection)*

A nostalgic look at

GLASGOW TRAMS

Since 1950

Graham Twidale and R. F. Mack

Silver Link Publishing Ltd

CONTENTS

AUTHOR'S ACKNOWLEDGEMENTS

THE publication of this book is the culmination of a lengthy project, but sadly my co-author, Bob Mack, who provided most of the black and white photographs seen in these pages, did not live to see the finished product. Bob was very well known in transport circles and his passing was a sad loss. I would like to express my most sincere thanks to Mrs Pamela Mack for ensuring that this book was completed - I'm certain Bob would have been as proud of it as I am!

Also, I would particularly like to record my thanks to Glasgow journalist Tom Noble for his very welcome assistance in the production of this book. His help, both in producing the 'present' pictures and also in carrying out caption research in Glasgow's Mitchell Library, were very much appreciated. I would also like to thank Bob Downham, now of Stranraer, and Jack Timmins, of Glasgow, for their help in research and caption writing. My thanks also to well-known Glasgow historian and author the late Jack House for his specialist assistance. I was also grateful for the help of Ian Stewart, whose knowledge of Glasgow's tramways is probably unsurpassed.

Any historical book always requires the assistance of a library at some stage, and I would like to pay tribute to the staff of Glasgow's Mitchell Library, who provided a wealth of useful information. At the 1988 Garden Festival site (formerly Princes Dock, see page 40) I was fortunate to enlist the support and assistance of Commercial Director Nigel Lindsay and also Transport Manager Gordon Baker, who arranged a pre-opening visit to the Festival tramway to enable Tom Noble to take the photograph required to bring the Glasgow tram story up to date. For making excellent new prints from the historic negatives, credit is due to John Fozard of Baildon, Yorkshire.

Finally I would like to thank my wife Kendal for her tolerance while I cluttered the dining room for weeks on end with books, notes and photographs, and my two sons, Matthew and Alastair, for their good-natured patience during the long evenings and weekends I spent 'locked away' reading and writing about Glasgow's marvellous trams during the compilation of this book.

Graham Twidale

First published in April 1988
Reprinted July 1988
Reprinted May 1989
Reprinted July 1990
Reprinted July 1991
Reprinted with minor revisions October 1994

British Library Cataloguing in Publication Data

A catalogue record for this book is available from the British Library.

ISBN 1 85794 049 0

Silver Link Publishing Ltd
Unit 5
Home Farm Close
Church Street
Wadenhoe
Peterborough PE8 5TE
Tel/fax (0832) 720440

Printed and bound in Great Britain

Left: A magnificent image of Glasgow City Centre in the late 1950s. 'Standard' car No. 73 is waiting at the Argyle Street junction traffic lights in August 1959; at the time the *Scottish Daily Express* was offering a new Daimler Majestic or £2,500 in cash as a competition prize. Note the Ford Zephyr, whose tail 'fins' and whitewall tyres illustrate the American influence on new car design of the time.

TRAMLINES
A Foreword by
JIMMIE MACGREGOR

I was in London when the last tram ran in Glasgow. I couldn't get away, and the sense of deprivation was profound. I also felt deeply ashamed, as though I had failed to attend the funeral of a much-loved granny. These feelings are common to people who have had a relationship with the Glasgow tram. A three-legged camel was more comfortable, and the inside of a rolling barrel filled with scrap iron would probably have been quieter, but we all loved the trams. They were to streamlined travel what the tenements were to gracious living, but no Glaswegian will ever admit to hating tenement life or trams!

People spoke to you on trams. It was useless to try to avoid this. Wee drunk men were always offering to "Gie ye a song, son." Folk offered you chips and girls smiled at you from the tail of the eye. Tram seats had real upholstery and there was lots of mahogany and brass. Some of the older cars had stained glass windows. Honest. The conductresses who had been recruited to the jobs in wartime tended to stay on, and conductors were a rarity. A favourite put-down of an uppity conductor was: " Away, ya big jessie, and gie the lassie back her job!" Tramcars were good for cosy mellow, yellow light on a dark night; steamed up windows, the smell of wet raincoats, and good patter. "How long will the next car be?"

"About 45 feet, haw, haw"

"Oh aye. And will there be a monkey conducting that wan as well? "Or: "Any seats up the sterr?"

"Aye, but they've a' got bums on them." A glance through this book will further stir the memory of anyone who has experienced the Glasgow tram, and it's a safe bet that its pages will be dampened by many a nostalgic tear.

Jimmie Macgregor.

INTRODUCTION

THE construction of a working tramway at the Glasgow Garden Festival site, using vintage tramcars, is an evocative reminder of the extensive system which served this city for 90 years, from August 1872 to September 2 1962, when the last of the Corporation's cars made their final journeys on Service 9 from Dalmuir West to Auchenshuggle . Glasgow's tramway system is still fondly remembered today, as well it might be, for it was the most extensive tramway in Britain outside London; its longest route (from Renfrew Ferry to Milngavie) was an astonishing 22 miles in length, although this was curtailed during the Second World War, in 1943, when long routes were cut back as an economy measure. The Glasgow Corporation tram

system at its peak was an enormous organisation; in 1923 (following the absorption of the Paisley and Airdrie systems) the transport department employed nearly 10,000 people, owned more than 1,000 trams using in excess of 200 miles of track, and served a population of 1,250,000 people.

I should stress the important point that this book is not intended to be a definitive history of the Glasgow tramway system, nor is it specifically concerned with technical information about the trams themselves. No, all this has already been covered authoritatively and professionally on previous occasions, principally by Ian Stewart, of the Scottish Tramway & Transport Society, whose book *The Glasgow Tramcar* is a superb

Above: A classic image of Glasgow in the 1950s. In sunny weather in April 1959, 'Standard' car No. 751 and 'Coronation' No. 1243 are pictured in characteristic surroundings at Glasgow Cross, where 751 is about to reverse on the crossover whilst working back to Shettleston. Note the driver, presumably off duty, riding 'on the cushions' at the rear of No. 1243. As with many old photographs, the background becomes more interesting as the years pass. Note the small Police telephone box, mounted on the short post on the pavement, visible to the right of 1243. These small telephone boxes, which replaced the large blue kiosks made famous by 'Dr Who', have of course since been superseded by modern personal radios. The granite setts, once a common feature of many cities and towns, disappeared progressively through the 1960s in favour of tarmac. *(FP002)*

This page: A fascinating comparison of the changing scene in the city centre, at Glasgow Cross. Above: A sunny day in April 1960 as 'Coronation' car No. 1195 passes beneath the Saltmarket/High Street trolley bus wires at Glasgow Cross station. Above the door of the imposing station building is an advertisement for Barr's famous *Irn Bru*, a soft drink manufactured since 1901 and still very popular today. There have been many myths and tales about *Irn Bru*, which has been believed to be a cure for hangovers, or that it can make the hair turn red! Made to a closely-guarded secret recipe, the 'ferric content' of the drink is at the heart of its mystique; one popular piece of folklore is that Glasgow taxi-drivers should not be allowed to put their bottles of *Irn Bru* near the meter, as it can affect the fare! Originally named *Iron Brew*, the abbreviated *Irn Bru* name was introduced in 1946 as the company thought the term 'Brew' was misleading, as the drink is not fermented or brewed in any way. Barr's, makers of *Irn Bru*, was founded in Falkirk more than 150 years ago by Robert Fulton Barr. The modern view (right) reveals that all trace of trams, trolley buses and trains have been swept away from Glasgow Cross, pictured here in March 1988. The fine buildings in the background survive, but the railway station, part of the Glasgow Central (Low Level) suburban network, was demolished following closure on October 5 1964. This line actually reopened as an electrified passenger route in 1979, but Glasgow Cross did not reopen. *(FP003)*

publication and highly recommended. *A nostalgic look at Glasgow Trams since 1950* is a nostalgic look back at the trams as part of the city scene between 1950, when rationing was still in force, and 1962, when the 'swinging sixties' were under way, and many aspects of life which had been familiar for generations were being swept away. In an age of cheap oil, the electric tram surrendered to the motor bus, whilst on our railways the steam locomotive capitulated to diesel and electric traction, after more than a century of faithful service.

The withdrawal of tram services in Glasgow in September 1962 was indeed a symbolic event, for with the notable exception of Blackpool and Fleetwood, whose trams still run today, the Glasgow trams were Britain's last and their disappearance truly marked the end of an era. Change was indeed a watchword in 1962, and the official conception of progress in these years produced some major casualties. In 1962, British Railways started modernising Euston, the London terminus of Glasgow's West Coast Main Line railway, and in so doing demolished the spectacular and historic Doric Arch fronting the station. Despite a huge protest, spearheaded by John Betjeman, the demolition went ahead, like the scrapping of Glasgow's trams, in the name of progress. Today, the demolition of fine buildings like the Doric Arch would never be permitted, whilst in cities like Manchester, plans are afoot to reintroduce trams! In European cities trams have always been important, but in the 'wind of change' sweeping Britain in the late 1950s and 1960s, and the desire for modernisation at almost any cost, they could not survive. The 1950s was indeed a 'watershed' decade during which many aspects of everyday life started to change irrevocably, and this book is a reminder of Glasgow in the years when streets were surfaced with granite 'setts' and everyone seemed to wear a hat!

Unless otherwise credited, all the historic black and white photographs in these pages come from the collection of the late Bob Mack, whose vast archive of more than

260,000 negatives covered many aspects of transport. Bob was very enthusiastic about this book but, sadly, he is no longer with us, having passed away in January 1987. It is a great personal sadness that he did not live to see this book in print, and I would like to express my thanks to Mrs Pamela Mack for making sure the project went ahead as her husband had wished. Bob was well known in transport circles, and photographs of buses, trams and trains from his huge archive have appeared in many publications. He gave us free access to his collection and the majority of photographs appearing in this book are published here for the first time. Many were taken by Bob himself, some by George Staddon, with others from collections acquired by Bob over the years. Choosing which pictures to include was a difficult job and I have selected those which I believe represent all that was best about the tramway system in its latter years, when the Kingston Bridge and the Clydeside Expressway were but planners dreams!

Glasgow has changed dramatically since September 4 1962, when in a Grand Procession following the end of public services, the trams finally trundled off into the history books, and into the scrapyard. The pictures in this book portray many buildings which the march of progress has changed, almost beyond recognition. Study the scenes carefully, look at the advertisements on the trams and streetside hoardings, the then-new motor cars (remember the trusty Austin A40 and Ford *Popular?*) - also, and this was a point which struck both myself and publisher Nigel Harris most forcibly -- the almost total absence of litter and rubbish in the neatly-swept streets. This contrasts strongly with the fast-food, throw-away attitudes we see so much of today.

Notice too, how many of the tram crews wore full uniform with pride -- how many bus drivers do you see today wearing a peaked cap? Remember also that these were the days before smokeless zones, and most homes enjoyed open coal fires, although many people, even then, were less than enthusiastic about the

daily chore of cleaning out the ashes and laying a fresh fire! Also, pity the poor coalman, struggling up flights of stairs with fuel for the bunkers of residents on the upper floors! The absence of smoke control did have its drawbacks of course, and helped blacken the city's fine buildings, many of which have in recent years been painstakingly cleaned. We are, of course, fortunate that so many have survived to be cleaned in the 1980s, for in many ways the 1960s was a decade of change, simply for the sake of change, and some fine buildings were razed to the ground in the name of progress, to make way for characterless glass towers. For many people, the hallmark of Glasgow is its distinctive tenement buildings, and you will see many examples in these pages, including those 'up market' buildings complete with half-tiled 'wally close.'

Overall scenes have also changed radically in recent years; in the city centre Charing Cross, St George's Cross, Anderston Cross, Parliamentary Road and Castle Street have been transformed, whilst other thoroughfares, such as parts of Sauchiehall Street and Argyle Street have been pedestrianised. Outside the central area, the construction of expressways, flyovers and other new roads have caused many buildings to be swept away. In the suburbs, change has been generally less dramatic and Hillhead, the area around the University, Hyndland and Pollokshields remain fairly timeless, for example.

This book is divided into five principal chapters and it has been inevitable that the dividing lines we have chosen are loose and rather arbitrary, and made purely for convenience in laying out the book. Each chapter examines the relevant areas individually north and south of the River Clyde, and in general terms, examines the city from East to West. One important point; the term inner city has in recent times become rather misconceived in its meaning and indicative of deprivation and decay. In the context of this book, this should be ignored!

To highlight the degree of change, both physically and in character, we have given 'past and present' treatment to some locations. Glasgow journalist Tom Noble assisted by re-photographing today some of the scenes from precisely the same spot as the old pictures (or as closely as possible) - and the comparisons are quite fascinating, for in some views, the surprises come as a result of what has survived, rather than what has gone!

So, there you have it. This book is unashamedly pure nostalgia for the pre micro-chip age. It is aimed at those with a fond affection for things that have been, and can never be quite the same, ever again. Hopefully, the pictures in these pages will revive memories of the squeal of wheelflanges, the hiss of the current collector on the overhead wire, and the staccato beat of wheels on railjoints as the cars clattered over the many junctions in the city centre, or raced along Mosspark Boulevard. The trams are still recalled with affection by many people, and I hope this book gives as much pleasure to its readers as it has given me in compiling it and finally, look closely at the many people pictured riding on the trams, doing their shopping and going about their daily business. You might just find yourself, or someone you know, captured forever on film in a bygone age!

Graham Twidale
March, 1988.

A classic view of Glasgow city centre which has changed little in general terms, compared with other parts of the city, though of course the trams are long gone, the buildings have been cleaned and the 'Capstan' cigarette advertisement has been removed from the facade of Central station. The tunnel-like bridge carrying the railway over Argyle Street has always been known as the 'Hielandmans Umbrella' and in this view 'Coronation' car No. 1199 is heading a line of three trams standing at the traffic lights at the junction of Argyle Street and Union Street. The 'Coronation' cars were introduced following successful trials with prototype No. 1141, which entered service in January 1937. Built at a cost of £4,406, No. 1141 was an immediate success and a second example, No. 1142, emerged from Coplawhill works on March 11 1937. Painted in spectacular silver-grey, red and blue livery to commemorate the Coronation in May 1937 of King George VI, this striking livery earned the nickname 'Coronations' for the entire original series of 152 cars which followed these two pioneers into service. Introduced at a time of increasing road traffic, the 'Coronations' featured trafficators (the arrow shaped lights beneath the drivers window), stop lights (the circular light at the bottom left corner of the front panel) and a rear view mirror (visible in this picture) for the driver. *(FP004)*

A BRIEF HISTORY OF GLASGOW CORPORATION TRAMWAYS

GLASGOW'S first trams, albeit of the horse-drawn variety, took to the city streets on August 19 1872, operated by the grandly named Glasgow Tramway & Omnibus Company. The pioneer route linked St George's Cross and Eglinton Toll (a distance of 2 miles 2 furlongs 2 yards), running via Cambridge Street, Sauchiehall Street and Renfield Street. By the end of 1872, new routes included: Bridgeton-Candleriggs (1m); St George's Cross-Belhaven Terrace (1m 2f 39yds); Crescents-Cambridge Street (7f 116yds); Bridge Street-Paisley Road (6f 40yds); Whiteinch-Crescents (2m 4f 20yds) and the St Vincent Place Branch (1f 20yds). The system expanded steadily through the 1870s and 1880s until by 1894 the total mileage was 31m 3f 55yds

The passage of the General Tramway Bill through Parliament in 1870 had 'paved the way' for the arrival of trams in Glasgow, but for the first few years the new Company suffered financial difficulties as a result of stringent conditions imposed by the City Corporation, which granted a lease for use of the streets by the Tramway Company. It is of interest at this point to note that the gauge between the rails was set at the unusual figure of 4ft 7 3/4in. This was because the Vale of Clyde track, between Fairfield shipyard and Govan Cross, had originally been laid as a goods line, to link the shipyard with the national railway network. Consequently, this track was laid at 4ft 7 3/4in gauge, so that trams could run conventionally on the rail top, whilst also allowing railway wagons to run on their flanges, in the groove of the rail; thus, all Glasgow's tram track had to be laid at the same non-standard gauge. This caused problems when trams from other systems were acquired, such as when the Liverpool cars were bought in 1953/4 and re-gauging was required.

As the years passed, the Company overcame its early financial troubles and the undertaking operated profitably. However, relations with the city fathers were generally strained and in 1894 the Corporation exercised its right, enshrined in the Act of 1870, to terminate its lease to the Company and assumed municipal control of the tramway system. This was not achieved without much animosity and the ousted and much-disgruntled tram Company offered no assistance whatsoever, and even succeeded in preventing the Corporation from acquiring its fleet of trams. A new fleet of vehicles was immediately required and between 1894 and 1898, the Corporation sub-contracted the construction of 384 horse-drawn trams. They reputedly cost £143 each and were generally accepted as being the ultimate for the time in modern lightweight transport. Car No. 543, housed in the Glasgow Museum of Transport, is a surviving example of the first type of tram owned by the Corporation. At the time of the takeover (which passed smoothly and was popular with passengers, the 31-miles of horse-drawn system was served by nine depots and nearly 3,000 horses.

By this time, other local authorities in the United Kingdom were introducing electric tramcars, Blackpool having constructed the country's first electric street tramway in 1885. Glasgow Corporation realised that conversion to electric power was the logical and progressive way forward (rather than steam traction , which was also considered!) and the decision to proceed with this major project was made in 1897. On October 13 1898, the first experimental electric service was introduced between Mitchell Street and Springburn; this proved highly satisfactory and no time was lost in electrifying the entire system. The first 21 new electric cars were completed in 1898, these being single-deck vehicles popularly known as 'room and kitchen' cars. No. 672 survives today in the Glasgow Museum of Transport, now housed in the city's Kelvin Hall. These cars were divided into two saloons with a centre loading platform (hence their nickname) and smokers used an unglazed portion, which was nevertheless fitted with protective blinds for use in poor weather! Some 120 horse cars were also converted to electric operation and remained in service until the 1919-1923 period, when they were replaced by new 'Standard' cars.

The first of the four-wheeled 'Standard' vehicles appeared in 1898/9 and they ultimately became familiar in all parts of the system. More than 1,000 examples were built, many surviving in service until the 1950s, with the final car not being withdrawn until 1961. A double-bogie variation of the 'Standard' design, built between 1927 and 1929, earned the nickname 'Kilmarnock Bogies' amongst enthusiasts. The bogie cars were designed to give a smoother ride, but they initially at least displayed a tendency to derail on tightly-curved city centre junctions, and they were thereafter generally seen on the comparatively straight 'red' routes along Argyle Street and London Road, spending most of their working lives on services 9 (Auchenshuggle-Dalmuir West) and 26 (Burnside-Scotstoun).

It is of interest to note that service numbering did not come into general use in Glasgow until 1938, although experiments in this field had taken place before the First World War. In the early years of the century, passengers were more accustomed to the colour-coded bands applied to the trams, denoting different routes. Colours initially used were:

Red: Anderston Cross-Whitevale; Glassford Street-Maryhill; Hillhead-Paisley Road; Shawlands-St Vincent Street.
Blue: Braehead Street-Crosshill; Govanhill-Stockwell Street; Paisley Road-Rockvilla.
White: Dalmarnock-Finnieston; Dennistoun-St Vincent Place;

Springburn-Mitchell Street; St George's Cross-Crosshill.
Green: Kelvinside-St Vincent Place; Paisley Road-Parkhead; Pollokshaws-St Vincent Street; Queen Street-Whiteinch.
Yellow: London Road-Queen Street; Overnewton-Paisley Road; Partick-Queen Street.

The total cost of electrification was £500,000 and by 1902, about 500 electric tramcars were in service, comprised both of new vehicles and also converted horse-drawn cars. Electricity was generated by the new power station, at Pinkston, which fulfilled this role to the last days of tramcar operation.

By 1910 almost 100 miles of track were in use, of which around one-third was outside the city boundary. At this time, congestion in the city centre was becoming a major

problem, but the tram service and network was comprehensive, the undertaking was performing well financially, the system was the envy of many other municipal authorities and by 1914 nearly 1,000 cars were in service. During the First World War, more than 3,000 tram crews left Glasgow to fight in France, and, sadly, one in six never came home. To ensure the trams kept running during the war, women were recruited not only as conductresses, but also as drivers. In general terms the system weathered the war fairly well and in

1922 the tramway's 50th anniversary celebrations were staged.

During the 1920s, major important arguments about the role of street tramways raised questions which four decades later were to result in the extinction of the public tram not only in Glasgow, but also (with the exception of Blackpool and Fleetwood) throughout the United Kingdom by the early 1960s. The costs of laying and maintaining track both in busy streets and rural areas, the problem of trams occupying busy streets and causing interruptions to

Above: An attractive rural picture which provides a strong contrast with the bustle of the city centre tramway routes. In April 1959, 'Standard' car No. 147 is pictured at Arden, en route back to Kelvingrove. Driving a tram on such a route, in sunny weather like this, must have been a pleasurable experience, and it is interesting to note that service No. 14 (a 'blue' service) was at one time the longest tram route in the United Kingdom, linking Milngavie with Renfrew Ferry - a distance of 22 miles! It was said that crews working this 'blue' service managed only two 'round trips' per shift! However, during wartime restrictions imposed in 1943, this service was cut back to run between Milngavie or Maryhill to Spiersbridge. In its final form (as seen here) this service linked Arden and Kelvingrove, prior to complete withdrawal on November 1 1959, when motorbus service No. 57 took over. More than 1,000 'Standard' cars were built in various guises between 1898 and 1924, with many later-built examples surviving to the last days of tram working. No. 147 was a 'Phase III' tram, introduced as a 'yellow' car in November 1913 and withdrawn after 46 years service in June 1959. *(FP006)*

traffic flow and whether more flexibly operated buses were a better alternative to trams were all discussed with increasing frequency, and from now they would never be far away. Despite this, the trams continued to serve the city and its people well, but in April 1932 a tram route was withdrawn at Paisley. It was the start of a process which ended at Auchenshuggle in September 1962.

Notwithstanding this sad development, the 1930s saw much progress on the tramway system and a landmark of particular note was the introduction in 1936/37 of the double-decker experimental saloon cars Nos. 1141 and 1142, forerunners of the highly acclaimed 'Coronation' cars. Class pioneer No. 1141 was seen on Albert Drive, undergoing official inspection, prior to service,

on January 1 1937, whilst No. 1142 entered traffic on March 16 1937. To say the least, these cars were quite revolutionary compared with their predecessors, and No. 1142 was especially impressive, for it emerged from Coplawhill Works carrying a livery of red, silver-grey and blue, to celebrate the Coronation of King George VI. Glasgow was to host the Empire Exhibition in 1938 at Bellahouston Park and more of the new cars were urgently needed in time for the opening on May 3 that year. By the end of 1938, almost 100 cars had been completed, and had become known as 'Coronations'. They were immediately popular with passengers and crews alike: they were fast, comfortable, heated, fitted with modern lighting and folding doors to minimise draughts.

By 1938, the Corporation had

finalised its route numbering system, though the colours were still retained, as shown in the accompanying table.

By July 1941, there were 152 'Coronation' cars in service, running alongside the rather elderly (certainly in terms of appearance) 'Standard' cars. The war years witnessed a lack of maintenance and great strain and in the Spring of 1941 there was a major air raid. Glasgow was an important industrial centre which made an essential contribution to the war effort, and the Corporation trams certainly 'did their bit' - in 1944 the fleet of cars carried 13,956,000 passengers over more than one million track miles, and collected more than £86,000 in fares. During the war years the colour code system began to break down, partly because the new 'Coronation' cars did not carry it. Thus, the 'white' codes were the first to go between 1938 and 1942, the other colours disappearing over subsequent years until they vanished completely by 1952, after which the route numbers became much more important. The effect of the war was felt for many years after 1945 and women tram crews served until final abandonment. A further 100 new cars of a Mk II 'Coronation' design were built between 1948 and 1952, these vehicles becoming known as 'Cunarder' cars. To the layman they were not dissimilar to the 'Coronation' predecessors, but there were detail differences both in bogie design and interior layout. After 1949, route numbers encompassed from 1 to 40 (with some gaps) and services were subsequently withdrawn, merged or replaced either by motor bus or trolley bus services.

By 1953 more cars were needed to replace older members of the ageing fleet, and 46 'Green Goddess' trams were bought from Liverpool. In 1954, the last trams built in the Corporation's own workshops were placed in service; these were officially designated as replacement Mk 1 'Coronations' for cars destroyed in a depot fire, the insurance money from which financed construction. Ex-Liverpool bogies were used for these 1954 vehicles, which could thus not be classed as 'new'. Of the six

SERVICE NO. & COLOUR	SERVICE
1/Green	Knightswood/Kelvinside-Airdrie
1a/Green	Dalmuir West/Scotstounhill-Springfield Road
2/White	Provanmill-Polmadie
3/White	University-Mosspark via Eglinton Street
3a/White	University-Mosspark via Paisley Road West
4/Blue	Keppochhill Road-Renfrew/Porterfield Road
4a/Blue	Springburn-Linthouse
4b/Blue	Lambhill-Linthouse
5/Yellow	Clarkston-Kirklee
5a/Yellow	Langside-Jordanhill
6/Blue	Riddrie/Alexandra Park-Scotstoun/Dalmuir West
7/Yellow	Millerston/Riddrie-Craigton Road
8/Red	Bishopbriggs/Springburn/Millerston/Riddrie/Alexandra Park-Newlands/Giffnock/Roukenglen
9/Red	Dalmuir West-London Road/Auchenshuggle
9a/Red	Scotstoun-Burnside
9b/Red	Scotstoun-Rutherglen
10/Blue	Rutherglen-Kirklee
11/Red	Maryhill/Gairbraid Avenue-Sinclair Drive
12/Yellow	Mount Florida-Paisley Road Toll
13/Red	Mount Florida-Hillfoot/Milngavie
14/Blue	Milngavie/Hillfoot/Maryhill-Renfrew Ferry
15/Green	Airdrie-Paisley/Ferguslie Mills
15a/Green	Uddingston/Tollcross-Paisley/Ferguslie Mills
16/Green	Whiteinch-Keppochhill Road
17/Red	Cambuslang-Anniesland
18/White	Springburn-Rutherglen/Burnside
19/White	Springburn-Netherlee
20/No Colour	Yoker/Clydebank-Duntocher
21/Bus Green	Provanmill-Crookston

examples built, four were later to become victims of the disastrous fire which destroyed Dalmarnock tram depot on March 22 1961.

Although there had been minor route closures before the Second World War, namely the short Finnieston branch, the Paisley-Abbotsinch route and the Kilbarchan line beyond Elderslie, the huge system remained virtually intact until the first major abandonment in 1949, when on April 3 that year Service No. 2, from Provanmill to Polmadie was converted to trolleybus operation , motor buses having covered the route since February 20.

The decision to abandon the whole system was first made in 1955, which was also a year of major decision in other transport undertakings, for in the same year the young British Railways organisation, formed at Nationalisation in 1948, announced its own Modernisation Plan, under which the steam locomotive would be banished from its tracks. This aim was achieved in August 1968, six years after the last tram had left Auchenshuggle. In 1955, following the closure of London's system in

1952, Glasgow Corporation took over the role of operating the largest tramway undertaking in the country, comprising 1,027 cars, 28 services, 11 depots and no less than 256 route miles. However, in these years, the street tramway throughout Britain was attracting criticism of increasing severity, and Glasgow was no exception. By the early 1950s, around 75 per cent of the trams were more than 30 years old, congestion was reaching intolerable levels and whilst the trams were not entirely to blame, they certainly exacerbated the problem, and also, revenue was falling short of that required to meet costs - and with each round of fare increases some passengers stepped off the tramcar platforms for the last time. By 1961/2, passenger totals had fallen to 9,731,000 (four million passengers less each year than in 1944) although this raised fare revenue of £185,200 -- inflation and rising prices are not new!

In 1956, the Corporation abandoned all tram services beyond the city boundary and as a result, Airdrie, Barrhead, Cambuslang and Milngavie lost their cars. In the following year,

Below: 'Standard' car No. 10 (built in 1923) leaves the rails for the last time at Baillieston and is jacked into the cradle of a James Connell's low loader for its journey to the scrapyard. *(FP007)*

Right: The sad end of 'Standard' car No. 291, partly dismantled in September 1958, at Connell's scrapyard at Coatbridge. *(FP008)*

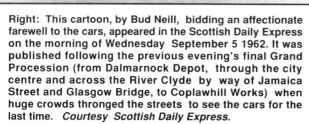

Right: This cartoon, by Bud Neill, bidding an affectionate farewell to the cars, appeared in the Scottish Daily Express on the morning of Wednesday September 5 1962. It was published following the previous evening's final Grand Procession (from Dalmarnock Depot, through the city centre and across the River Clyde by way of Jamaica Street and Glasgow Bridge, to Coplawhill Works) when huge crowds thronged the streets to see the cars for the last time. *Courtesy Scottish Daily Express.*

tram services were withdrawn from Paisley. During the mid-1950s, it was estimated that Glasgow's last trams would be withdrawn circa 1970. By 1958 the end was forseen as coming in 1963, but both estimates were incorrect for as already reported, the last cars ran in September 1962. The last services to survive were:

WITHDRAWN IN 1960: Nos. 1,3,10,23 and 30
WITHDRAWN IN 1961: Nos. 16,18 and 29.
WITHDRAWN IN 1962: Nos. 9,15 and 26.

Doubtless some people were glad to see the trams go, but many others were sorry to see them disappear and certainly, the final services were crowded by city dwellers bidding farewell to the cars which had served them so well, for so long. Also, the final procession was an emotional affair. A number of Glasgow's tramcars are preserved, both in this country and overseas, thereby providing the opportunity to cast a nostalgic glance back to the days when hundreds of trams served Glasgow. In September 1962 we all believed that the city had seen its last tram. However, the creation of a tramway at the Garden Festival in Govan Road, on the site of the former Princes Dock, using preserved cars, has enabled the writing of another Chapter in the history of Glasgow trams. Back in 1962, we all thought that the words 'The End' had been written beneath the story of the cars. In 1988, more than a quarter of a century later, will the Garden Festival tramway enable the last chapter of this story to be told? We shall just have to wait and see!

THE CITY CENTRE

GLASGOW city centre was the hub of the City Corporation's tramway system, and the busy network of streets, and the bridges over the River Clyde, were host to a steel web of tram track. At city centre junctions, trams swerved sharply over tightly curved rails as they executed a 'right angle' turn from one street into another, or clattered noisily across the intersecting tracks. The noise and bustle of the trams provided a constant background of sound against which the city went about its business.

The highly complex nature of the city centre tramway system is clearly shown on the accompanying map It is a fascinating glimpse of an amazing web of rails and wires, and it reveals how some of the complex junctions were modified through the years. For the purposes of this book, the 'city centre' is an area whose boundaries roughly encompass Charing Cross, Cowcaddens, Glasgow Cross and Jamaica Bridge.

Above: Former Liverpool car No. 1035 (left) and 'Coronation' car No. 1225 pause alongside the 'Brighter Homes' home improvements shop in Trongate in April 1958. Both cars are in generally good condition, especially 1225, which appears to be 'ex-works', with very clean paintwork. No. 1035's destination blind windows are very grimy, and prospective passengers would have needed very sharp eyes to discern that it was heading for Maryhill, via Normal School! Car No. 1035 was one of a batch of 46 trams purchased from Liverpool Corporation in 1953/4; the last of these cars (No. 1056) entered service in Glasgow in May 1956. The Liverpool trams (built during the 1930s) were bought by Glasgow Corporation ostensibly to replace its own ageing 'Standard' cars, but they were longer by 2ft than the home-built 'Coronation' cars and the increased 'overhang' created clearance problems and after trials they were only able to run on two routes: No. 29 (Broomhouse-Milngavie, from November 3 1956 to Maryhill) and No. 15 (Baillieston-Anderston Cross. Several modifications were required before the Liverpool cars could run in Glasgow including regauging of the trucks from 4ft 8 1/2in to Glasgow's non-standard 4ft 7 3/4in. Known commonly in Glasgow as 'Green Goddesses,' the ex-Liverpool cars were not problem-free in Glasgow: motor and electrical faults were common during their early days in the city, and all 46 examples had been withdrawn by 1960. They were not over-popular with Glasgow crews and had a very short life in the city compared with the native cars. No. 1055 (built in 1936) did survive however, and is preserved today in its original guise as Liverpool car No. 869, at the National Tramway Museum, at Crich, in Derbyshire.

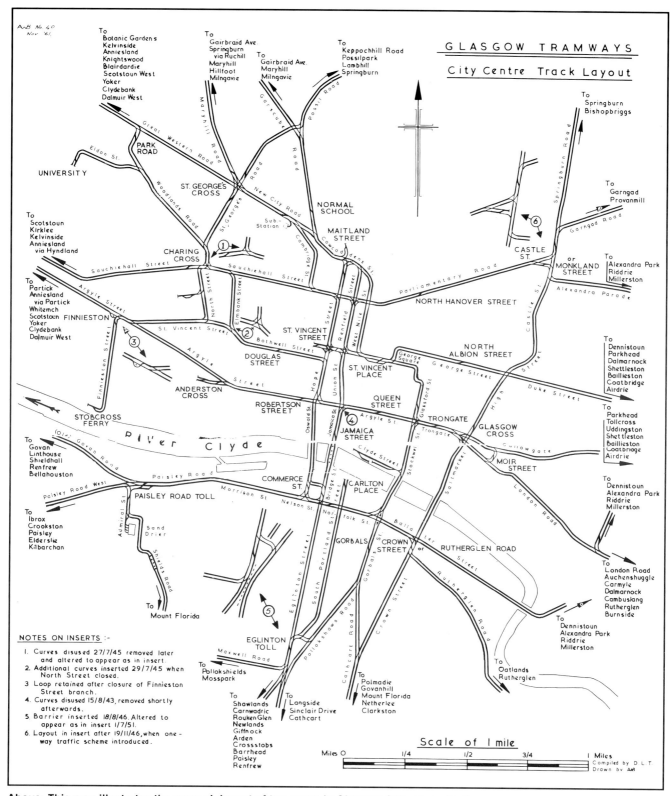

NOTES ON INSERTS :-

1. Curves disused 27/7/45 removed later and altered to appear as in insert.
2. Additional curves inserted 29/7/45 when North Street closed.
3. Loop retained after closure of Finnieston Street branch.
4. Curves disused 15/8/43, removed shortly afterwards.
5. Barrier inserted 18/8/46. Altered to appear as in insert 1/7/51.
6. Layout in insert after 19/11/46, when one-way traffic scheme introduced.

Scale of 1 mile

Miles 0   1/4   1/2   3/4   1 Miles

Compiled by D.L.T.
Drawn by A.B

Above: This map illustrates the general layout of tramways in Glasgow City centre in the years after 1938. Detail alterations (such as the removal of crossovers) were implemented at various times and precise dates are not given here. The Finnieston Street tramway closed in 1927. Names shown in capitals appeared on car destination screens. The map shows clearly the intricate nature of the city centre tram system both north and south of the River Clyde and the congestion experienced on these busy streets as road traffic increased during the first half of the 20th century (see overleaf, lower picture) was one of the chief weapons in the armoury of the anti-tram lobby.
Scottish Tramway & Transport Society, Courtesy A.W.Brotchie & D.L.Thomson.

Right: 'Cunarder' car No. 1356's front panel shows signs of a recent collision as it runs along Gallowgate, passing an advertisement for Players cigarettes which depicts a smiling young couple. In these years, cigarette smoking was very much more common than it is today, when the health hazards are more widely appreciated. Gallowgate is said to be a derivative of 'Gallowgait'; gait meaning 'way to,' hence 'way to the gallows.' *(FP010)*

Left: An evocative view of London Road, at the junction of Moir Street, in August 1958, as 'Standard' car No. 987 clatters over the points en route to Shawfield. Note the hoarding on the gable end above No. 987: the *Empire News* was promoting coverage of Mrs Manuel's story of her son Peter, the multiple murderer from Uddingston. He stood trial in May 1958 accused of eight murders (two triple murders and two separate killings of two girls) and although acquitted on one charge owing to lack of evidence, he was convicted on the other seven counts and was hanged at Barlinnie prison on July 11 1958. *(FP011)*

Right: A typically busy scene in Argyle Street in April 1960, with 'Standard' car No. 64 and 'Coronation' car No. 1168 illustrating how the trams attracted criticism for adding to street congestion as increasing numbers of cars, vans and lorries thronged the city streets. Commercial traffic such as the Tennent's dray lorry (visible to the right of No. 1168) which needed to park at the kerbside to unload, combined with rapidly growing private car ownership in an age of cheap oil, to create an environment in which the conventional street tramway could not survive. The trams were pollution-free and in today's comparative terms would be much cheaper to run than diesel-engined buses; however, in 1962 they were regarded as an obstruction and simply 'had to go.' Lewis's department store is visible in the background of this scene and on the right (beyond the scaffolding) is a sign advertising the site for the new Marks & Spencers store, On the extreme right of the photograph can be seen a shop sign proclaiming the availability of kilts. *(FP012)*

Left: Argyle Street, looking east during April 1960, as viewed from the upper deck of a tram. Pedestrianisation was unheard of as 'Cunarder' car No. 1345 ambled away from the photographer, towards Burnside. The 'Cunarders,' introduced in 1948, were a Mk II version of the highly successful pre-war 'Coronation' cars and the first example, No. 1293, entered service in December 1948. A total of 100 'Cunarders' were built in the Corporation workshops at Coplawhill, the last to emerge being No. 1392, in February 1952. Production of the new cars was slow, due partly to a shortage of space and partly to the works' ongoing workload in maintaining the ever-ageing fleet of 'Standard' cars. Pressure of work at Coplawhill was so intense at one stage that coachbuilders R.Y. Pickering Ltd, of Wishaw, were asked to tender for the partial construction of tram bodies. However, the combination of staff recruitment at Coplawhill with new working practices prevented the need for sub-contracting, and all 'Cunarders' were completed 'in-house' by the Corporation workshops. Two examples survive today. (FP013)

Above, left: The 'Hielandman's umbrella' and Central station is the setting for 'Coronation' car No. 1191, turning from Argyle Street into Hope Street, in April 1960. The lines in the foreground, which formerly carried routes Nos. 4, 22, 27 and 40 over George V bridge into the south side of the city, were disused by this time. At least four other cars are visible in the distance; today, a plethora of buses would greet the eye at this location. Glasgow Central station is still busy and the passenger concourse has been refurbished to a high standard, whilst retaining its characteristic Caledonian Railway wooden panelling. In our opinion, the station is vastly superior in atmosphere to its rather clinical counterpart at Euston, at the south end of BR's West Coast Main Line. (FP014)

Above, right: 'Standard' car No. 17 follows 'Cunarder' No. 1362 along Hope Street in August 1958, both cars working service 22 to Lambhill. This was a very busy junction, with an almost endless procession of trams clattering over the points and crossings. This area has seen comparatively little change over the years, and Central Hotel (right) opened in 1883, still dominates the scene. In bygone days, when many celebrities travelled by train to Glasgow, they invariably stayed here. It is recalled that the great scientist Einstein, stepping off his train, could not believe the huge crowd was there to greet him. He was correct, for he soon saw the more obvious candidates who had travelled on his train - Laurel and Hardy! (FP015)

Right: A sunny day on Glasgow Bridge, or Jamaica Bridge as it is usually known, in April 1959 with 'Standard' car No. 459 following replacement Coronation' car No. 1398 over the River Clyde and into the city centre. No. 1398 was one of a batch of 'replacement Coronation' cars built following the destruction of five Mk 1 'Coronations' in a fire at Newlands Depot on April 11 1948. Although not strictly speaking an 'all-new' car, No. 1398 was the last tram to enter service in Glasgow, on October 7 1954 , but it ran for a mere eight years before withdrawal in September 1962, whereas No. 459, which entered traffic in December 1903, worked for nearly 56 years before withdrawal in June 1959. More than 1,000 'Standard' cars were built in Glasgow, in four basic sub-classes, in the 63 years following the introduction of pioneer 'Standard' car No.686, in 1898. *(FP016)*

Above: A very interesting view of the approach to Jamaica Bridge in April 1958. 'Standard' car No. 29 is passing Broomielaw (on the left) and approaching the bridge (immediately behind the photographer) en route to Giffnock, flanked by a 'split-screen' Morris Minor saloon of which few survive today. Assuming the clock on the Paisley's building is working, the picture was taken during the late morning, showing bustling commercial activity along Jamaica Street. Car No. 29 entered service in March 1922 as a 'red service' tram and it worked for more than 36 years before being withdrawn and scrapped in June 1958. *(FP017)*

Left: A busy day in Renfield Street. 'Standard' No. 502 is disembarking passengers at the junction of Bath Street. The picture was taken in hot, sunny weather and the lady passenger on the upper deck has taken advantage of the central drop window above the destination screen to 'take the air ' as she enjoys a cigarette and a chat with her companion. This drop-window was provided to give the crew access to the rope attached to the bow collector, from which the tram drew its electric current. This rope was used to reverse the bow collector when cars reached a 'dead-end' terminus. The tall building in the background housed the municipal transport offices, prior to their transfer to more modern premises in West George Street. *(FP018)*

Above, right: 'Coronation' car No. 1292 (the last-numbered example in the Mk 1 series built 1936-1941) swings round the sharply-curved junction from Hope Street into Bothwell Street in April 1961, on its way to Springburn. Although originally envisaged as a domestic shopping area, Bothwell Street was taken over more by offices and car showrooms. 'Cunarder' No. 1364 is approaching the photographer as it proceeds along Hope Street. The Abbey National Building Society still occupies the impressive corner building, and still uses the silhouette of a couple walking beneath an umbrella as its company logo. Also of interest is that this building, in common with many others in the city centre, still carry their tram wire support fixtures (rosettes) in the 1980s, as shown by the modern picture (above, left) which shows the corner of Bothwell Street and Hope Street in March 1988. More than a quarter of a century after the trams ended their careers, reminders of their presence can still be seen around the city. In other city thoroughfares - for example, London Road, Auchenshuggle - the cross-wire support posts were still *in situ* in early 1988. The cars only carried advertisements after 1949, and the 'Heinz Beans' example carried by No. 1292 promotes a product which is indeed still familiar today! *(FP019)*

Right: There are clear indications here of the independence of the Glasgow pedestrian: note the old lady crossing Hope Street in fearful odds as 'Coronation' cars 1263 and 1211 approach from one direction, and an Austin A35 van from the other! Car No. 1263's side panel advertisement reads: 'Many 'Bulletin' features now in the Scottish Daily Mail,' whilst on No. 1263 we learn that the *Evening Citizen* was still in print. The final edition of this newspaper was published on March 29 1974. Glasgow's trams provided excellent value for money for the passenger. After the Second World War, rising costs established a cycle of rising prices, but in the immediate pre-war years the maximum day time fare was just 2 1/2d, for which the longest trip available was from Milngavie to Renfrew Ferry - a distance of 22 miles!

Above: A scene which has changed completely today: this is Hope Street, at the junction of Cowcaddens Street and Maitland Street. It's April 1961 and well-filled 'Coronation' car No. 1241, working service 29 from Maryhill to Tollcross, is about to swing left to run along Hope Street, from where it will join Argyle Street at Central station. The motor vehicles, which include a Ford *Popular* saloon, a Vauxhall *Cresta*, an Austin *Cambridge* and a Leyland *Comet* lorry are all worthy of note and would be collectors items themselves today. In the background, the Clydesdale's store was having a sale, and in those days kitchen floors were usually covered with Linoleum, which in more recent times has been superseded by a variety of vinyl products. Clydesdale is still one of Scotland's most famous retailers, but the company has had a rather chequered history. Founded in 1911 by Isaac A Levin, the Clydesdale Supplies Company first traded on Glasgow's famous 'Barras' market, selling carrier bags, oil cans, bicycles and furniture. Steady growth throughout the 1920s and 1930s meant that the company was poised for major expansion after the Second World War, and the demand for radios and phonograms in the early 1950s provided Clydesdale with the ideal market, which was successfully exploited. Expansion was a rapid and the stores sold furniture, radios, phonograms, carpets and television sets - and the company even introduced its own brand radio and phonogram under the name 'Rolland.' However, in 1961 Isaac Levin died, leaving his son Teddy to take over the business: by 1962 the company was bankrupt and, as a result of heavy debt, was sold for just one penny! Acquired first by bankers Lloyds & Scottish, the company was subsequently the subject of a management buy-out, and since 1985 the company has been under Dutch control. Clydesdale today is best known as an electrical retailer, but its operations also include a credit company and an electrical repair service.

(FP021)

This page: A fascinating comparison indeed. The older view (above) depicts 'Coronation' car No. 1187, crossing the junction of Cambridge Street (where a pair of tracks diverge to the left), Dalhousie Street and Shamrock Street, and hurrying down New City Road, towards St George's Cross. The picture was taken in April 1959, and in the intervening years change at this location has been dramatic. New City Road, once a busy thoroughfare, is now a *cul de sac* at this point, a result of the construction of the M8 motorway, which now cuts across the scene, as illustrated in the modern view, taken on March 17 1988. Back in 1959, the Glasgow Savings Bank (subsequently absorbed by the Trustee Savings Bank) occupied the ground floor of the impressive domed building and whilst the TSB closed this branch on May 8 1987, close scrutiny of the modern view reveals that the bank's three-circled emblem can still be seen on the wall to the left of the gated door. At the time of going to press, the upper floors were still used for residential purposes, but the building was being offered for sale.   (FP022)

Right: 'Coronation' car No. 1147 is pictured in April 1959 on the north side of George Square, with the magnificent arched roof of Queen Street station in the background. This was one of the late Poet Laureate John Betjeman's favourite Glasgow scenes, which is unfortunately today concealed by modern development - principally an extension of the North British Hotel. On the front of the car, the *Evening Citizen's* advertisement poses the question: 'Did the brass hats send an army to die in Flanders Fields?' This is probably still a question worthy of debate today, but equally probably, will in all likelihood never be satisfactorily answered. Car No. 1147 held the dubious distinction of being the last Glasgow tram to be scrapped, on February 9 1963. This picture shows particularly clearly the 'stop' light, trafficators and rear view mirrors fitted to the 'Coronation' cars.   *(FP023)*

Above: 'Standard' car No. 18 turns from Renfield Street into St Vincent Street in May 1958, with the impressive, although smoke-blackened columns of the Bank of Scotland providing an imposing background. Note the small single 'traffic light' on the tram wire support pole on the right - an early version of the 'Pelican' pedestrian crossing so familiar today. This picture illustrates the final form of the Glasgow 'Standard' car, built in various guises between 1898 and 1924.

Earlier versions had featured fully- open top decks and vestibules, or partially covered tops with open end balconies, as represented today by preserved car No. 779. Some examples worked for more than 50 years - giving far greater service than could ever be expected from a modern bus! The cars were always popular with drivers, who liked their spacious driving conditions. In their later years the bodywork was often in a parlous condition, and one of the authors vividly

recalls a fast run down Maryhill Road on one of these old cars, on service 29, where it appeared that the top deck was about to part company with the rest of the tram! The 'Standard' cars soldiered on in ever dwindling numbers until mid-1961, when they became extinct. Fortunately, some examples are preserved, including No. 1088, the last-built example of 1924. This car, withdrawn for preservation in June 1961, is displayed in the city's Museum of Transport, at the Kelvin Hall.   *(FP024)*

Left: All the buildings seen here in Bothwell Street have now gone. Nobody would seriously question the need for constructive progress, but it is sad that spectacular Victorian buildings such as the YMCA, seen here towering in the background, should have been demolished. 'Standard' car No. 368 is waiting as car 36 uses the crossover whilst on a short working of service 18, back to Ruchill. The picture was taken in May 1958, when the 'Standard 8' motor car on the right was popular . Note the advertisement on car No. 368, which exhorts the public to 'Go Gay at Butlins Holiday Camps'! *(FP025)*

Right: The same viewpoint as the picture shown above, photographed in February 1988. Everything has changed except the general pattern of the streets and buildings. The trams are long gone, the tenement block and YMCA have given way to modern buildings of brick and concrete and the stone setts have disappeared beneath a tarmac surface. 'Standard' cars 36 and 368 illustrate the durability of Glasgow's home-built trams: No. 36, which entered service in in February 1922 worked for more than 37 years before withdrawal in June 1959. No. 368 had an ever better record: entering service in May 1910, it was not withdrawn until October 1958, after 48 years in traffic.

Left: Car No. 1005, pictured in Bothwell Street in April 1960, was a unique tram with an interesting story. Construction started in 1946, although the car did not enter service until December 1947, when the striking non-standard livery of three shades of blue (lined in gold, black and white) soon earned the car the nickname 'the Blue Devil'. No. 1005 was also unique in being designed as an experimental one-way car; passengers were supposed to board at the front and disembark at the rear. Consequently, the car could not reverse at terminii and started its career at Possilpark Depot to work the circular service (No. 33) to Springburn, but was later transferred to Maryhill Depot. However, the car caused much confusion with passengers, who were accustomed to boarding at the rear, and in 1949 No. 1005 was modified with a rear entrance and front exit. The standard livery was also applied during 1949 and in a subsequent conversion (in 1956) a single exit/entrance was provided. No. 1005 was withdrawn in 1962. Finally, note the route indicator reads '81' instead of '18' - perhaps the conductor was suffering from 'the morning after the night before!' *(FP026S)*

A nostalgic look at
GLASGOW TRAMS
Since 1950

Right: A sunny day in Argyle Street in July 1955. The gentleman leaning against the tram stop pole is apparently deep in thought, as 'Coronation' car No. 1145 passes, working service 15, which at that time linked Airdrie and Anderston Cross. In earlier days, this had been a 'green' service (see page 12). The car on the left is one of the 46 ex-Liverpool cars acquired by the City Corporation transport department in 1953/4; they were known in Glasgow as 'Green Goddesses' in recognition of their original Merseyside livery.

Below: An evocative scene at the junction of Sauchiehall Street and Hope Street, also in July 1955. The tram running from left to right across the junction is heading east along Sauchiehall Street, whilst 'Standard' car 491 (left) is about to move off, away from the camera, along Hope Street. The traffic lights on the left are just changing from red and amber to green, so the other car must have passed the west-east traffic lights as they were switching back from green to red! *Both: C. N. Banks Collection*

Left: A busy Sunday morning scene in Argyle Street, still in July 1955, with 'Standard' cars Nos. 325 and 626, bound for Burnside and Moir Street respectively. In the left background, adjacent to the Grant Arms restaurant, garden lounge and grill, is Glasgow Central Station and the famous 'Hielandmans Umbrella'.
C. N. Banks Collection

Left: Passing Gale & Barclay's Ford showroom in April 1960 is 'Cunarder' car No. 1315, turning from Elmbank Street into St Vincent Street, en route to Scotstoun. In the showroom, new *Consul* and *Zephyr* models were on display whilst outside, the two gentlemen on the corner reveal the popularity of the trilby hat. Not only has this scene itself changed beyond recognition since this picture was taken, but white-coated policemen on point duty are seldom seen nowadays! *(FP027)*

Right, upper: In April 1960, 'Coronation' car No. 1235 turns from Sauchiehall Street into Renfield Street, on a short working on service 23 from Maryhill to Dennistoun Depot. The newly asphalted surface of lower Sauchiehall Street can clearly be seen, following the removal of trams from this part of the street the previous year. This part of Sauchiehall Street is now pedestrianised. In the distance can be seen the distinctive structure of the Copeland & Lye department store, now alas no more, this building having been demolished in 1973/4. *(FP028)*

Right, lower: An evocative scene at the junction of Sauchiehall Street and Elmbank Street in April 1960 as 'Replacement Coronation' car No. 1395 leads 'Standard' car No. 121 past the premises of Gray & Co., monumental sculptors. No. 1395 is en route to Mosspark, via the smart suburb of Pollokshields, whilst No. 121 is waiting to swing right into Elmbank Street on Route 10 (to London Road) which will cross service 3's path again at the junction of Argyle Street and Jamaica Street. In the left background can be seen the Locarno Club and Ballroom, and also the Grand Hotel at Charing Cross, now alas no more; this building was erased from the landscape during motorway construction in 1969. *(FP029)*

INNER CITY SERVICES

Above: Tram crews enjoy an unexpected break on a sunny day in Springfield Road in May 1958. 'Standard' car No. 300 has 'jumped the rails' on the crossover, blocking both tracks, and awaiting the arrival of assistance. The incident appears to have happened recently, for chaos usually followed within minutes of obstructions of this nature, as other traffic formed a queue, awaiting clearance of the stricken tram. At this time, car No. 300 was in its 49th year of service, but in cosmetic terms at least appears to be in good condition. Introduced into service as a 'red' route car in May 1909, the car was scrapped in September 1960, after 51 years in traffic *(FP030)*

Left: In April 1959, 'Standard' car No. 184 at Bridgeton Cross, whilst working a Shawfield service. The serried ranks of chimneys on the tenement rooftops indicate why the city's buildings presented such a blackened appearance in these years! The introduction of smokeless zones and systematic cleaning of buildings has transformed Glasgow in recent times *(FP031)*

Above: During its last month of service, 'Standard' car No. 824 climbs the gradient in London Road, in April 1959, bound for Shawfield. Introduced into service in August 1900, No. 824 was therefore withdrawn and scrapped just four months before completing its 60th year of service - a worthwhile record indeed! On the right are the *Oxford Tavern* and the James Boardman Stamp & Brand Works, whose dustbins are standing on the pavement, awaiting the arrival of the Corporation dustcart. In the distance stands St Andrews House, a Salvation Army Hostel.   *(FP033)*

Right: A most familiar sight to Glasgow tram passengers - the interior of a "Standard" car, complete with bare light bulbs! Note also the leather straps attached to the ceiling handrails, provided for the benefit of standing passengers. The longitudinal side seats, with their individually contoured backs, were very comfortable indeed, much more so than their counterparts on modern buses!

GLASGOW TRAMS SINCE 1950

Below: The junction of Duke Street and Cumbernauld Road in May 1958, viewed from the upper deck of a tram, as 'Standard' No. 53 rounds the curve on its way to Riddrie. Service 7 (Millerston/Riddrie-Bellahouston (Jura Street) was a lengthy journey, running via Bellgrove Street and Abercromby Street, Bridgeton Cross, James Street, Kings Bridge, Ballater Street, Paisley Road Toll and Govan. The service was taken over by trolley bus on June 14 1958, and in this view the pairs of overhead wires needed for trolleybus operation are already in place. Paton Street (the home of Dennistoun tram depot) joins Duke Street adjacent to the Beatties bread advertisement. The small corner shop was an outfitters named Elspeth, which specialised in children's wear, ladies skirts and the hire of wedding dresses. The modern view of this location (lower picture) reveals that the small shop still stands and whilst it appears to be abandoned today, close inspection of the original print shows that part of the 'Elspeth' sign above the window can still be discerned. Otherwise, there has been much change. The setts have long since given way to tarmac, whilst industrial development now occupies the skyline. Buildings on the right-hand side of Duke Street have been demolished, but the tenements beyond the junction with Cumbernauld Road have been cleaned and modernised. In front of this building, the street lamps seen in the older view are still in place. *(FP032)*

Above: A sunny day in August 1960 finds car No. 64 standing outside the depot in Paton Street, Dennistoun, also the home of Beattie's bakery. You can almost smell the freshly-baked bread! *(FP034)*

Right: A scene which has changed almost beyond recognition, for everything in this picture has gone, swept away to make way for the network of motorway bridges, flyovers and link roads which now occupy this location. The picture shows a bustling scene at Townhead, at the junction of Castle Street and Parliamentary Road in September 1959. 'Coronation' No. 1249 is approaching the photographer and joining Castle Street on service 6 to Alexandra Park; note the truncated tracks which formerly served Routes 25, 32 and 33 to either Springburn or Bishopbriggs. The trolleybus wires visible here had been in place since 1949, when service 2 (Polmadie-Provanmill, via Crown Street, Saltmarket, High Street and Garngad Road) had been replaced. The bus heading away from the camera is a Glasgow Corporation Leyland *Titan* (with Alexander bodywork) of 1958 whilst a Scottish Bus Group Bristol *Lodekka* is overtaking a Glasgow Corporation Daimler/Weyman bus on service 37. *(FP035)*

Right: Everything in this picture has gone and the site is now occupied by modern roads. In May 1959, 'Cunarder' No. 1306 turns into Monkland Street on service No 6 to Scotstoun. Sprinting past on the right is six-wheeled trolleybus No. TD27, a 1950-built vehicle. Glasgow's trolleybus system was the last new example of its kind opened in Britain, and it lasted just 18 years, from 1949 to 1967. The total fleet of 174 double and 21 single deck trolley buses operated on six routes: Nos, 101, 102, 105, 106, 107, and 108. They were known in Glasgow as 'the silent death!' *(FP036S)*

Left, upper: A relatively quiet spell in Sauchiehall Street in August 1959, as 'Coronation' No. 1153 and 'Standard' car No. 103 meet, with Charing Cross in the background. Note the pennant-shaped flags in the middle distance; these were hung in connection with the Scottish Industries Exhibition taking place at the Kelvin Hall. The small advert on the car's leading dash panel is promoting the Scotland versus Poland boxing international, staged at at the Kelvin Hall on September 17 1959. At this event, Olympic lightweight champion Dick McTaggart was beaten by Kazimierz Pazdior, on a majority points decision. Overall, Scotland lost six bouts to four and there was uproar when police had to escort Polish referee Adam Cwiklinski out of the Hall when he 'counted out' light heavyweight Alf Barker before the count of 10, in the first round. The referee was banned from further bouts. *(FP037)*

Left, lower: On a lovely summer's evening in August 1959, 'Coronation' No. 1228 passes the bowling greens of Kelvingrove Park; its next stop will be the Art Galleries, which houses one of the finest municipal collections in Britain. On a tranquil evening such as this, the alternately approaching and receding 'hum and rumble' of the cars as they passed was a highly distinctive aspect of Glasgow life. The design of Glasgow's 'Coronation' cars was a triumph for the Corporation team, for not only was prototype car No. 1141 a revolutionary step forward, it was also significant in being a success from the outset. There are usually many changes required before a prototype can be multiplied into a successful class, but the series which followed No. 1141 (introduced in January 1937) were all-but identical in essential respects. The 'Coronations' were pleasing to the eye, and comfortable for both passengers and crew - many drivers accustomed to the open platforms of the old' Standards' were amazed to find a separate driving cabin - with a seat! Not all aspects of the original design were incorporated in the whole fleet. One less-than-successful facility fitted to 1141 was a switch to reduced road noise in the saloon by closing a vent. This was provided because No. 1141 was fitted with loudspeakers, to enable the driver to announce the 'stops.' However, this laudable idea was quietly dropped after passengers had been entertained to some choice remarks, aimed by a tram driver at some other vehicle, after a collision was only narrowly avoided at Eglinton Toll! *(FP038)*

Right: At the junction of Sauchiehall Street and Argyle Street, 'Standard' car No. 168 and 'Coronation' No. 1183 pass en route to Anniesland and Burnside respectively. Near this location, the Kelvin Hall (formerly an exhibition centre and circus venue) now houses the Glasgow Museum of Transport, which has been transferred from its previous home in the former paintshop of the tram works at Coplawhill, in Albert Drive. Note the old-style road signs and bollards on the right and, above the door of Clark's restaurant, the *Coca Cola* logo, still in use today. It is the only aspect of the picture which does not appear dated in any way today. *(FP039)*

Above: A picture with plenty of detail to examine - including the man with his arm in a sling! 'Kilmarnock bogie' No. 1099 is rumbling along Dumbarton Road, near the Kelvin Hall and the Art Galleries - the pinnacles of which can be discerned slightly to the right of the tram's bow collector. No. 1099 is working service 9 (Dalmuir West to Auchenshuggle) destined to be Glasgow's last tram service. The Ford 'Thames' van on the right is in gleaming condition, and the cyclist behind the Austin 'Somerset' must have been in a great hurry - quite apart from the rough ride given by the setts (rather than the tarmac alongside) he is travelling perilously close to the motor vehicle! Viewed from the top deck of a car travelling in the opposite direction, the scene also includes a Leyland/Alexander 1959-built bus (alongside the kerb) from Old Kilpatrick Depot. It is working the long service 32 from Glasgow to Balloch, on the banks of Loch Lomond. This was a busy service which suffered heavily from competition from the railway in the early 1960s following introduction of electric 'Blue Trains' on the Balloch-Helensburgh-Airdrie corridor, which offered a better frequency of service and faster journey times. Nevertheless, the bus service survived and, since bus de-regulation in 1986, this portion of the service has been redesignated as No. 5/5A -- a six-minute frequency cross-city link from Old Kilpatrick (or Faifley) to Easterhouse, operated by ex-London Transport *Routemaster* buses, some of which are older than the vehicle seen here! *(FP040)*

Left: Hayburn Street (in which the cars are standing) was the home of Partick Depot and the tracks outside D.M. Hoey's shop rumbled to the constant procession of cars starting and ending their duties. Here we see 'Coronation' No. 1189 (on the right) about to enter busy Dumbarton Road, to take up duties on service 16, which ran between Scotstoun and Keppochhill Road, whilst 'Cunarder' 1306 heads for the depot. Hats, caps, collars and ties are amongst the wares on offer in Hoey's shop, which is still there today, selling mens and boys clothing. Hoey's was founded in 1898 by Donald McColl Hoey, who bought this particular shop in 1924. The company also had a big branch at St George's Cross (see page 34) which was its headquarters until closure in 1969. Hoey's then moved its offices to premises in Maryhill Road, and then (in 1981) to Victoria Road, which is the company's base today. Company takeovers have become very common today, so it is pleasant to report that Hoey's is still a family firm, for D.M. Hoey, grandson of the company's founder, is the Managing Director. Note the Austin A40 saloon and the Triumph *Herald* - once very common, but relatively rare today. *(FP041S)*

Above: A picture which is highly evocative of the general atmosphere of cities and towns all over the country during the 1950s. The intricate web of rails forming the complex junction in the foreground, and the large expanse of setts (or 'cobbles' as they were called in other parts of the country!) is an impressive sight indeed. The noise made by the cars as they clattered across these junctions was as much a part of Glasgow life as shipbuilding and railway engineering. In this view, 'Standard' car No. 999 is about to leave Maryhill Road and head across the junction into St George's Road, bound for Shawfield. The pre-war Leyland *Titan TD5* bus owned by Walter Alexander & Son is working one of the busy routes from the Milngavie direction. The buildings on the street corner survive, in refurbished condition, as do the Gentlemen's public conveniences, reached via the steps behind the railings. *(FP042)*

Above: A bustling scene at 5.10pm, during the evening 'rush hour', at St George's Cross during August 1958. 'Replacement Coronation' No. 1393 is waiting at the traffic lights, bound for Merrylee. As indicated by the mass of rails in the foreground, this was an extremely busy tramway junction, and comparison with modern view of this location (right) in February 1988, is an interesting exercise. The *Empress* theatre was originally designed by Harry McElvie as an opera house for West Enders, who would thus not have to travel into the city centre for their entertainment. However, this dream was not realised and the theatre has had a chequered career. Jimmy Logan bought it for £80,000 in May 1964 and renamed it the *Metropole*. However, the building has been abandoned since the early 1970s and in November 1987 permission was granted to demolish the theatre to make way for new development. At the time of going to press, the old Empress stood derelict and abandoned, a monument, perhaps, to things that might have been. *(FP043S)*

Right: In April 1960, 'Standard' No. 75 heads away from the camera at the junction of Maryhill Road and Bilsland Drive, bound for Maryhill, as 'Coronation' car No. 1232 approaches en route for Shawfield, from Springburn. One wonders if the young man on the upper deck of car 75 has lost something, whilst on the far street corner, a group of young ladies are attracted by the display in the *Cumfy Feet* shop window. The Citroen saloon, registered G1, is also of interest. It is said that this registration was originally the property of the Glasgow representative of the Automobile Association. Thus, when the time arrived for the Lord Provost of Glasgow to have an official car, the registration which might normally have been chosen was not available; thus 'GO' was chosen, and is still in use today. Back in 1958, the driver of the Citroen cannot have dreamed what value his car number would have, 30 years into the future! *(FP044)*

Left: A pleasant memory of trams in Great Western Road. In sunny weather during August 1959, 'Standard' car No. 181 draws to a halt to allow passengers to disembark close to Kirklee Terrace, as a uniformed motor cyclist approaches on the nearside. Note that the rider is wearing a soft peaked cap - this was in the days before the wearing of crash helmets was made compulsory. In the background is the city's Botanic Gardens, where some of the world's finest orchids are grown. This car was bound for Blairdardie, served by a tramway extension from Knightswood Cross, opened on July 7 1949. The small advert carried by the car is interesting: the Scottish Daily Express was promoting its Great Silverstone Contest, with a prize of either £3,000 or a new car. Assuming the car was of comparable value to the cash prize, it must have been a luxury model indeed, at 1959 prices! *(FP045)*

Right: Some fine examples of smart West End mansions can be seen in this pleasant view of car No. 673, entering Great Western Road from Hyndland Road in May 1958. Note that the once universal setts were no longer being replaced following street repairs, and that tarmac was being used to 'patch' the surface: Once the trams were gone, setts gave way to tarmac at an increasing rate. Adverts carried by the tram promote Crown wallpapers, still popular today, and the *Evening Times'* golf tips, by Ben Hogan *(FP046)*

Left: Paisley Road Toll in May 1958, as 'Standard' car 643 approaches the end of Admiral Street, bound for Govan Depot. This tram was extensively rebuilt in 1947, following serious damage in a street accident; it was eventually withdrawn and scrapped in June 1959. On the skyline in the left distance can be discerned the tower at Glasgow University, whilst to the left of the Imperial building is the British Railways General Terminus, a goods station. In 1957 this station had been re-equipped to handle iron ore to Lanarkshire Steelworks, but was superseded in 1980 by the more modern facilities at Hunterston. (FP047S)

Right: A latter-day view of the same location at Paisley Road Toll, during February 1988. The Imperial building is undergoing renovation, but the dockside crane has disappeared and the site of the General terminus is now occupied by new development. On the left, the 'Old Toll Bar' is still in business, with the same sign in place above the window, although the stonework above looks better after thorough cleaning, to remove decades of soot and grime.

Left: A characteristic view of old Glasgow in Allison Street, during a quiet spell in May 1958. Melvin Motors, who in those days specialised in Humber and Hillman models, are still in business, although not at this location. Today these names, and that of Rootes, have been consigned to the history books, together with the trams. Here we see car No. 409, pictured from the upper deck of a tram travelling on the other track, working on service 12, a busy cross-city route, especially on 'big game' days at Hampden Park. Mount Florida was the southern terminus of this route. Car No. 409, was built in 1905 as a 'red' route car, for the Anderston Cross-Whitevale, Glassford Street Maryhill, Hillhead-Paisley Road and Shawlands St Vincent Street services. It was rebuilt in 1946 and was finally withdrawn in February 1959. (FP048)

Below: An attractive broadside view of 'Standard' car No. 643, about to rattle over the junction at the convergence of Pollokshaws Road, Nithsdale Street and Allison Street in May 1957. The car is operating service 12, which never entered the city centre, running as it did from Mount Florida to Paisley Road Toll, or on to Linthouse. and Shieldhall. The advertisement for Bisto has withstood the passage of time remarkably well. The hoardings on the left include adverts for Guinness (as revealed by mention of the famous Toucan) and Stergene. *(FP049)*

Left: The graceful support poles are an attractive feature of this view of car No. 334 in Shields Road in May 1957. The car was working service 12, to Paisley Road Toll. These ornate overhead wire support poles were very unusual, and other than these, the only other similar examples on the whole system were to be found on Jamaica Bridge. Note also the longitudinal stone slabs laid in the cobbles, provided to make life easier for horses, labouring up the hill with heavy carts; the smooth stone presented much less resistance to solid wheels than the uneven setts. *(FP050)*

Below: Obstruction in Paisley Road. 'Standard' car No. 27 has drawn to a halt, as the driver is clearly uncertain as to sufficient clearance for his vehicle to pass the contractors Bedford tower wagon. One of the workmen is seemingly waving the car slowly past the lorry as his mates work on the overhead supply. The contractors appear to be in the early stages of fitting trolleybus wires, in readiness for the superseding of various tram routes. Paisley Road Toll can be seen in the background, with the angel standing proudly aloft on the roof of the former Ogg Brothers department store. *(FP051)*

Right: A 'birds nest' of overhead tram and trolleybus wires adds to the atmosphere at Harvie Street in September 1958, as car No. 751 passes en route to Mount Florida. The tracks diverging to the left led to Brand Street, home of Govan Depot. The advertisements on the trolleybus are promoting Premium Savings Bonds, which had been introduced just less than two years before. The first prize draw took place in June 1957, when the top prize was £1,000. Today, Premium Bonds are as popular as ever, but at the time of going to press the top monthly prize had reached the dizzy heights of £250,000 . The dock site in the background, indicated here by the row of crane jibs, is the site of the Garden Festival. *(FP052S)*

GLASGOW TRAMS SINCE 1950

SUBURBAN ROUTES

Below: The end of the line. On September 4 1962, the very last pair of cars to reverse at Auchenshuggle stand in the sunshine for the photographers and other well-wishers. The scheduled service had ended two days previously, but such was the public demand that a special farewell service was operated from Anderston Cross to Auchenshuggle, from September 2-4, with sixpenny fares. Car No. 1174 carries a banner on its leading upper window which proclaims: "Goodbye Trams - For All You've Been To Us - Thanks." This once-great system (only London's had exceeded it in size in the UK) had in around six years been whittled away to extinction.

The advertisement on No. 1174 dash panel reads: "Be a shareholder in Britain - Invest in National Savings" At that time, the National Savings Bank was based at Blythe Road, London; it subsequently moved north of the border and is now housed in Glasgow at Boydstone Road, Cowglen. *(FP053)*

Left: An attractive view of 'Kilmarnock bogie' cars Nos. 1108 and 1113 at Auchenshuggle terminus in July 1956. The side advert on car 1108 is interesting: The *Scottish Daily Express* was offering a prize of £10 a week for life -- which in 1956 must have seemed a small fortune. The 'Kilmarnock bogies' entered service in 1927/9 but early experience revealed a persistent tendency to derail at sharply curved junctions, and for much of their lives the cars were based at Partick and Dalmarnock depots for service on the relatively straight 'red' routes along London Road, Argyle Street and Dumbarton Road. Unkind critics have said that early tramways General Manager James Dalrymple invented the name Auchenshuggle so that people would travel there, if only as a result of curiosity! The name also has Gaelic origins, meaning 'field of rye' but despite this, the name Auchenshuggle does not appear at all on the modern *Geographia* map of Glasgow. *(FP054)*

Below: 'Standard' car No. 496 arrives at the terminus of service 17 in Cambuslang Road, Farme Cross in August 1958, three months before this service was withdrawn, without replacement. Standing awaiting departure time is 'Cunarder' No. 1373, which is in rather woebegone condition and has clearly been involved in a minor collision. It seems rather strange to see commercial air travel by BEA (British European Airways) advertised on the side of a vintage tram, built in 1903 ,when aviation was in its infancy. The 'overlap' between the last years of the tram period and the start of the so-called 'jet age' has produced a striking and thought-provoking image of the 1950s. At this time, Britain had two principal commercial airlines, the other being BOAC (British Overseas Airways Company) with which BEA was merged in September 1972, to become British Airways, now a privatised company. *(FP055)*

Right: Former Liverpool 'Green Goddess' No. 1006 stands in the sun at Baillieston terminus (Martin Crescent) in April 1959. No. 1006 was the first-numbered in the series of 46 such cars (Nos. 1006-1056) acquired in 1953/4 by Glasgow Corporation, to replace older cars of the 'Standard' design. This car survived for only six months after this picture was taken, it being withdrawn and scrapped in November 1959. These lightweight streamlined cars were bought from Liverpool for £500 each, including delivery and it is said that a miscalculation by Liverpool's Transport Department meant that as transport costs were higher than anticipated, the cars were more of a bargain than they might otherwise have been. However, the Liverpool cars were longer than the native Glasgow designs and presented operational problems as a result of excessive 'overhang' on sharp curves, which restricted their route availability. They had all been scrapped by 1960. *(FP056)*

42

Above, left: It's April 1959 in Baillieston Road, and 'Standard' car No. 324 is pictured at rest embarking passengers wishing to use service 15, to Anderston Cross. This is a so-called Phase II 'Standard', of which 121 examples were built at Coplawhill between 1904 and 1910.

No. 324 was introduced into service in December 1909 and worked until May 1959. *(FP057)*

Above, right: Hawthorn Street, Springburn in August 1958, and 'Standard' car No. No. 672 (in the distance) is about to use the

crossover and return to Burnside whilst sister car No. 108, in very clean condition, is working the service 33 'circular'. Listed for preservation, this car was destroyed in the disastrous fire at Dalmarnock Depot on March 22 1961, and its place was taken by car No. 22. *(FP058)*

Above: In rather work-stained condition, 'Coronation' car No. 1177 pauses on Elmvale Street siding, Springburn, in April 1958. The rather weary-looking horse eyes the camera suspiciously whilst 'nose-bagging 'the bread bin! Note that at this time, pools company Littlewoods were promoting a recent record win of £206,028, for a tuppeny stake! *(FP059)*

Above: An interesting array of advertisement hoardings flank the Forth & Clyde canal bridge in Possil Road in September 1959, as 'Coronation' car No. 1254 rumbles downgrade away from the camera. On the left of the bridge, Cadbury's were promoting their famous 'Roses' chocolates, not only still with us today, but also still packaged in the same distinctively shaped box. Over on the right we learn that *Bachelor* tipped cigarettes cost 3/4 for 20 (less than 17p) and that Murraymints are "too good to hurry mints!" Other advertisements include William Youngers beer, Campbell's soups and John Begg scotch whisky. *(FP060)*

Above: After a turn of duty on Route 29, 'Standard' car No. 75 is pictured in Celtic Street , about to enter Maryhill Depot. This street had the distinction of being the shortest street in Glasgow - it is barely longer than the car itself! *(FP061)*

Above: 'Coronation' 1150 near Maryhill Park in April 1958. Glasgow has over 70 parks, and the oldest public park in Europe; Glasgow Green was laid out from 1814. London's Royal parks are older, but were originally not open to the public. *(FP062)*

Left: An elevated view of 'Coronation' cars Nos. 1153 and 1284, pictured from the canal bridge in Maryhill Road in August 1959. Amongst the hoardings on the right is an advertisement for Billy Eckstine, who was appearing at the Glasgow Empire. Other posters promote Persil soap powder, Kellogg's famous Corn Flakes, Campbell's soups and McEwan's Export ale. *(FP063S)*

Above: With Lambhill in the distance, prototype 'Coronation' car No. 1141 passes the 'Vogue' cinema, Ruchill, in Balmore Road at the junction of Hawthorn Street and Bilsland Drive, in September 1958. The main feature at the Vogue was *The Big Country*, starring Gregory Peck, Jean Simmons, Charlton Heston, Carroll Baker and Burl Ives. The film has been described as: "...a big scale western with a few pretensions to say something about the Cold War, all very fluent, star-laden and easy to watch." The next week's offering was *The Gift of Love*, starring Lauren Bacall, Robert Stack and Lorne Greene. Nowadays it's a case of 'eyes down for a full house, for the films have given way to bingo at the 'Vogue.'

Following the introduction of No. 1141 in January 1937, a total of 152 similar cars had been built by 1940. No. 1141's original body was destroyed by fire at Newlands Depot in April 1948, but the car was subsequently rebuilt and survived until withdrawal for scrap on October 26 1961, Fortunately, four examples of the class survive. *(FP064)*

Right, upper: A gathering of cars in Elmvale Street, Springburn, during pleasant weather in April 1957. 'Standard' cars Nos. 244 (nearest the camera) and 831 (middle distance) are identifiable. No. 244 is working service No. 4, which at that time linked Springburn and Paisley North: in May 1957 the service was cut back to Hillington with the ending of all Paisley and Renfrew tram services. In the late 1950s, mass public X-ray sessions were common, and No. 244 is advertising this service. However, the 'Last week' wording added to the original advertisement has produced the rather alarming exhortation: "Last week to save your life, X-ray now." *(FP065)*

Right, lower: Springburn Road, near the Elmvale Street junction, in May 1959. Rather battered 'Standard' car No. 262, with only a week or so of life left at the time, is working a Bishopbriggs service. Springburn was a busy community and at this time the North British Locomotive Company's works, which had built steam locomotives for service all over the world, were still at work. However, the days of steam traction were numbered and chiefly due to the company's failure to move with the times, and also its heavy reliance on export orders, redundancies began in 1960 and liquidation followed in 1962. The bus is one of 100 Daimler vehicles (with Alexander bodywork) bought in 1957, and is seen working route 37 (Croftfoot or Castlemilk to Springburn).which replaced tram route 19 on February 2 1949, one of the first tram routes to be superseded by buses. *(FP066)*

Left: Passing Ruchill Hospital (right) is 'Standard' car No. 283, in fine, sunny weather in August 1960. The hospital remains open today. *(FP067)*

Below: Shirley Bassey, with a full supporting bill (including King's Sea Lions!) was appearing at the *Glasgow Empire*, whilst a season of Grand Opera was under way at the *Alhambra* in May 1957, as 'Cundarder' No. 1335 climbs away from the railway bridge in Clarence Drive, bound for Anniesland. The hoarding in the centre background reveals that the *Odeon* was screening *Robbery Under Arms*, starring Peter Finch, David McAllum and Jill Ireland. Critics were not impressed by the film, which was described thus: " A howlingly dull film version of a semi-classic adventure novel. A rambling story with no unity of viewpoint is saved only by excellent photography." *(FP068S)*

Left: 'Kilmarnock bogie' car No.1121, bound for Dalmarnock Depot, follows 'Coronation' No. 1228 (heading for Riddrie) along Dumbarton Road, Whiteinch, on a sunny evening in September 1959. No. 1121 was built by R.Y. Pickering Ltd, of Wishaw and worked from October 1928 until May 1960. The 'Kilmarnock bogie" cars earned their nickname because their trucks were built by the Kilmarnock Engineering Company Ltd.The Corporation normally built its own cars at Coplawhill, but in the late 1920s the works was busy modernising older 'Standard' cars and following the emergence of prototype bogie car No. 1090, in November 1927, construction was, sub-contracted to a variety of builders. Hurst Nelson of Motherwell built 30 cars, whilst R. Y. Pickering and Brush, of Loughborough, built 10 each. The fleet of 51 cars was in service by January 1929. *(FP069)*

Above: It's September 1959 and 'Coronation' cars Nos 1268 (nearest the camera) and 1266 stand at Scotstoun terminus (Balmoral Street), ready to work back to Alexandra Park and Keppochhil Road respectively. The 'Cunarder' crossing the junction in the distance is heading towards the city centre. (FP070)

Right: Swinging round the curve at the junction of Dumbreck Road and Nithsdale Road is 'Replacement Coronation' car 1394, carrying an advertisement which reads 'Electricity brings a modern home within your purse.' The gates of Bellahouston Park are in the background. (FP071S)

Top: The unmistakable location of Mosspark Boulevard in March 1960, as 'Coronation' No. 1242 passes Dumbreck Crossover whilst working on service 3, which ran from Mosspark to the University, via the splendid Victorian suburb of Pollokshields. This service ceased in June 1960, two months after this photograph was taken, and the modern scene (above, left) reveals that whilst the tree shapes remain basically unchanged, the tramway is long gone. (*Modern picture:Tom Noble*)   (*FP 072*)

Above, right: Passing the junction of Mosspark Boulevard and Paisley Road West in May 1958, is 'Standard' car No. 313. The severed metals in the foreground were used during the Empire Exhibition of 1938, held in Bellahouston Park.   (*FP073*)

Right: An evocative view of old Glasgow, during the glorious summer of 1959. Pollokshaws Road is pictured in July of that year, when summer fashions were to the fore as 'Replacement Coronation' car No. 1395 led a procession of motor vehicles as it restarted on its journey to Arden after pausing to offload passengers. This sort of traffic congestion provided the anti-tram lobby with the ammunition it needed to hasten the demise of the cars. The photographic company Ilford, still very active today, was taking the opportunity of the superb weather to promote its films : 'Be snap happy, buy Ilford' urges the large advertisement carried by No. 1395. The small poster on the dash proclaims: 'All Glasgow reads Cliff Hanley's column', in the Evening Citizen. *(FP074)*

Left: In August 1958, 'Standard' car No. 308 is pictured passing Princes Dock, where a ship is moored at the quayside. This part of Glasgow has changed much since this photograph was taken for the Princes Dock site was chosen as the location for the Glasgow Garden Festival. Back in 1958, such a prospect would have been judged inconceivable, for at this time, in addition to being an important industrial centre, the city was an important port and it was said that everything from sewing needles to railway locomotives made in Glasgow were shipped through the docks Shipping activity today is much reduced, and gone are the days when the New Year was greeted by a deafening chorus of ships sirens. *(FP075S)*

Left: It's September 1954, and in this view we see 'Coronation' car No. 1162 at Airdrie, about to return to Anderston Cross. Airdrie was one of the furthest outposts of the tramway system, and in 1938 service 15 must have been one of the longest on the system, linking Airdrie and Ferguslie Mills. Wartime restrictions witnessed the end of such long services and after 1943 service 15 ran as shown in this picture. Tram services were withdrawn from Airdrie in November 1956.
(FP076S)

Above: A pleasant view recalling Glasgow suburbia during the early 1960s. 'Coronation' car No. 1252 grinds slowly uphill in Stonelaw Road, towards the terminus at Burnside in April 1961. The car driver behind must simply be patient! The 'Coronation' heading downhill, bound for Springburn, is being followed by a works van from Hillington's Rolls Royce factory, still in business. Two months after this picture was taken, all tram services to Burnside were withdrawn, giving motor car drivers a clear run at the hill. (FP077)

Left: The rather uneven setts in Stonelaw Road, Rutherglen, will be giving the driver of the new Vauxhall *Victor* (on the left) a bumpy ride, as 'Standard' car No. 41 passes in the opposite direction, heading for Burnside. On the right, a short-trousered schoolboy turns to stare at the photographer at work: he doubtless couldn't understand quite why he should be taking pictures of such an everyday subject as a tramcar! We frequently fail to appreciate quite how rapidly the commonplace fixtures and fittings of today are destined to be the nostalgia of tomorrow. Even the black and white striped road signs, surmounted by the red reflective triangle, and once taken so utterly for granted, are now virtually extinct - and notice the small-wheeled wicker shopping trolley too! (FP078)

Right: An attractive broadside view of 'Standard' car No. 41 at Hillcroft Terrace, Colston, in May 1959, en route to Bishopbriggs. The picture highlights the extremely sharp curves of the short-lived terminus used by service 25, from October 1954 to March 1955. This was not a success and the service was re-extended to Bishopbriggs. (FP079)

Left: A fine evening scene during May 1958 at Bishopbriggs, as 'Standard' cars Nos. 45 and 84 turn into the Kenmure Avenue terminus, where they will reverse to embark on long cross-city journeys to Carnwadric and Crookston. There are two views generally on how Bishopbriggs came to be so named: one idea is that this was the location where the Archbishop of Glasgow resided in Catholic times, or that it is derived from the 'riggs' (farmland) of the Bishop. Whatever, Bishopbriggs has changed considerably since this photograph was taken and the A803 road to Kirkintilloch carries a far heavier volume of traffic than in the 1950s. (FP080).

Above: The trim suburb of Milngavie on a wet afternoon in 1955, with 'Standard' car 609 waiting to start the long journey back to Calderpark Zoo, at Broomhouse. *(FP081)*

Above: Glasgow Road, Clydebank, in May 1962, as 'Coronation' 1222 approaches the crossing of the railway to John Brown's shipyard, where many famous liners were built. *(FP082S)*

Above: Five cars. including 'Standard' No. 292 on the left at the head of the queue and 'Kilmarnock bogie' No. 1119 (on the right) gather at Dalmuir West in April 1958. For those of us who recall the cars, it seems difficult to believe that three decades have passed since this picture was taken. Its clarity and sharpness are such that you could believe it was taken yesterday! The picture clearly shows that whilst generally similar in overall appearance to the 'Standards, the 'Kilmarnock bogie' cars had distinct characteristics of their own. On the upper deck, the leading windows were smaller in size on the bogie cars than on the older 'Standards.' Guinness and Heinz baked beans were being advertised on the left, whilst on the opposite side of the road, the tree shows that it has recently been the subject of an extremely heavy pruning! *(FP083)*

Right: This view shows the city boundary at Thornliebank station, at the junction of Thornliebank Road and Boydstone Road, in May 1959. Against the background of the Hillpark housing scheme, 'Coronation' car No. 1217 moves over the junction, bound for Arden. The tracks diverging to the left in the foreground formed the Carnwadric extension, opened in August 1949 as the last tramway development of its kind in the United Kingdom. *(FP084)*

Above: The extensive nature of the Glasgow Corporation tramway system meant that its cars could be seen not only in city centre and industrial environments, but also in more rural surroundings, as illustrated by this April 1959 view in Thornliebank Road. 'Coronation' car No. 1258 is leading 'Standard' No. 73 en route to Rouken Glen and Arden respectively. The gentleman on the right has dismounted from his bicycle to chat to the two ladies, who are taking an interest in the photographer. There is not another vehicle in sight and note the complete absence of litter.
(FP085)

GLASGOW TRAMS SINCE 1950

Above: A delightful scene near Arden in April 1959. Rooks are nesting in the treetops in the distance as 'Standard' car No. 73 leaves Nitshill Road reservation and enters Thornliebank Road, Spiersbridge, watched by a group of pedestrians. Travelling by tram over such a route on a beautiful day such as this was indeed a memorable experience. *(FP086)*

Left: Paisley Cross, with Gilmour Street station in the background, is the setting for this picture, taken in April 1957. Gabardene coats were clearly very popular at this time! Paisley was a great weaving centre and its shawls and scarves are famous the world over. The town's independent tram services were operated by the Paisley District Tramways Company, until August 1 1923, when Glasgow Corporation took control. Most of the ex-Paisley cars had been withdrawn by 1953 but No. 1068 survived and is preserved by the National Tramway Museum, in Derbyshire. In 1988, it returned to Glasgow to work on the Garden Festival tramway. *(FP087S)*

Right, upper: An impressive sight in Porterfield Road, Renfrew in May 1957 as up to a dozen cars are lined up ready for the end of the days business at the neighbouring Babcock & Wilcock factory. This spur (off Paisley Road) was used to hold cars waiting to take workers home. Note that no form of 'stop block' is provided at the end of the track and extreme caution was needed by motormen! *(FP089)*

Right, lower: Cars Nos. 6 and 1004, two of five experimental lightweight four-wheeled cars, are pictured in Paisley Road at the junction of Porterfield Road, in Renfrew. Built between 1939 and 1941, these cars served for a large part of their lives (between 1951 and 1957) on the Glenfield-Renfrew Ferry services, working from Elderslie Depot. When the Paisley services ended in 1957, they were transferred to Govan and were used on shipyard 'extras' during 'rush hours'. All five cars were scrapped by Connell's of Coatbridge in 1959. Designed as a cheaper alternative to the 'Coronations' both to construct and maintain, the four-wheeled lightweight cars could be described as being only partially successful. Being non-standard, they were less popular with crews and maintenance staff, but some aspects of their design were incorporated into later-built cars, and in modifications to older vehicles. *(FP090S)*

DEPOTS AND WORKS

Above: A splendidly animated view of Coplawhill Works, in August 1959. At this time, the tramway had a further three years life ahead and here we see 'Cunarder' car No. 1310 being lifted clear of its bogies by the travelling overhead crane. In the foreground are an assortment of trucks undergoing overhaul, whilst in the centre of the shop floor an interesting conversation seems to be in progress - was it about work, or perhaps an inquest into the 'old firm' game last Saturday? *(FP091S)*

Above: A general interior view at Coplawhill Works, in Albert Drive, in August 1959, with a variety of 'Coronation' cars undergoing repair. All heavy repairs and modifications were undertaken at Coplawhill, where much of the fleet had also been designed and constructed. It was unusual for the Corporation to sub-contract construction, and examples such as the 'Kilmarnock bogie' cars were isolated exceptions. *(FP092)*

Right: The entrance to Govan Depot on a wet day in September 1958 with "Standards' Nos. 338 (left) and 16 awaiting their next turns of duty. Also at rest in the depot are a number of 'Cunarder' cars and trolley buses. At the beginning of the 1950s there were 11 depots serving the tramway system, at: Coatbridge (capacity 19 cars) Dalmarnock (119 cars), Dennistoun (134 cars), Elderslie (47 cars), Govan (see below), Langside (130 cars), Maryhill (93 cars), Newlands, (201 cars) Parkhead (80 cars), Partick (124 cars) and Possilpark (133 cars). The Govan Depot, in Brand Street, accommodated 128 cars, working services 4, 7, 12, 21, 22, 27 and 32. The Depot closed to trams on November 15 1958 but was used to store cars awaiting scrapping until February 28 1959. *(FP093)*

Top: The interior of Dalmarnock Depot in August 1959, with 'Standard' No. 681 and 'Cunarder' cars 1378 and 1383 stabled awaiting their next turns of duty. The stairs and walkways provided to allow access to the upper body panelling were constructed of wood, which provided much fuel for the flames which engulfed the depot in March 1961. Above, left: The tangled wreckage of twisted, blackened steel tells its own story after the 1961 fire in which 50 cars were destroyed. Barely identifiable in the left centre of the picture is 'Standard' car No. 108, which had been listed for preservation. Above, right: Despite the devastation, the roofless depot was used until abandonment of the system, in 1962. 'Standard' car 585 is seen in the entrance of the Depot during March 1962, probably whilst working an enthusiasts tour. *(FP094 - top picture)*

Right: From a small installation in Admiral Street, dried sand was provided for all the major depots. As with railway locomotives, tram drivers could apply a trickle of sand to the rails in wet conditions, thus improving 'grip' and preventing wheelslip, especially on gradients. Sand was transported from the drier to the depots in special works cars such as No. 39, pictured here outside Maryhill Depot. No. 39 was one of two cars specially built for this duty in 1939, and was provided with a low cab roof, so that the car could be driven into the sand drier building. No. 39 was scrapped in December 1959. *(FP095)*

Left: As with all tramway systems the majority of maintenance and engineering work on the track and overhead supply was done after dark, when the passenger service was over. Consequently, although Glasgow Corporation employed a good number of works cars, they were seldom seen on the streets during daylight hours, when they were normally stabled in the permanent way department base, adjacent to Coplawhill Works. Here we see works car No. 27, a tool van car used to transport equipment to the scene of an engineering job, and provide warmth and shelter during tea breaks - for these cars were also fitted with a coal stove. This car was originally built in 1904 as a Paisley District double deck car, in which form it served until withdrawal in February 1933, conversion in the form shown here following in January 1934. The car was eventually sold in October 1962, after 28 years of chiefly nocturnal service of an unspectacular, but nonetheless important nature. *(FP096S)*

Right: Elderslie Depot in August 1956, with 'Standard' car No. 258 in occupation, probably during an enthusiasts tour. This depot housed 47 cars, which at this time worked services 21, 28 and 32, in addition to local duties in Paisley. Elderslie Depot closed on May 11 1957, when its cars were transferred to Govan, Possilpark, Maryhill and Dalmarnock; the building was subsequently sold. No. 258 was introduced into service in June 1912 as a 'yellow' service car and it worked for nearly half a century before being withdrawn after 48 years service in June 1960. This was a very respectable working life, during which No. 258 doubtless travelled millions of miles and carried millions of passengers. *(FP097)*

Right: There's not a soul to be seen in the depot yard at Newlands in April 1959 when 'Standard' car No. 234 and a 'Cunarder' accompanied many other trams stabled after their day's work. The depot housed up to 200 trams, for services 3, 8, 14, 24 and 31. Newlands-based cars were transferred to either Dennistoun or Partick when the depot closed on October 21 1961, after which it was used as a motorbus garage. (FP098)

Left, upper: A familiar sight in Coplaw Street, as seen here at the junction of Annandale Street, in November 1958, was the School Car, used for training motormen. Allocated for many years to Langside Depot, it was frequently seen shuttling up and down a single track section (adjacent to the Samaritan Hospital) which was not used by normal services. Originally built in 1904 as a Paisley & District open-top double deck car, this vehicle became part of the Glasgow Corporation fleet in 1923, when the Paisley system was taken over. It was converted as a single deck car and adapted for use as a training vehicle in 1925, in which form it served until August 1960. With the end of the tramway system in sight, there was little need for a School Car and No. 1017 was withdrawn. (FP099)

Left, lower: Although not strictly a works car, this picture of a steeple-cab electric locomotive at work on the Corporation's system is appropriately included in this section. It was owned by the Fairfield shipyard for shuttling railway wagonloads of materials to and from the yard. As mentioned elsewhere in these pages, it was because of this quarter-mile link that the whole Glasgow network was built to the non-standard gauge of 4ft 7 3/4in, rather than 4ft 8 1/2in, for railway wagons have a different flange profile to tramcars, and by constructing the track to this unusual gauge, the rail wagons were able to run on their flanges, in the groove of the tram track, between the main line and the shipyard. This picture shows the scene at Govan Cross in September 1957. In today's terms, the thought of a railway train running through a city street is quite bizarre; back in 1958 it was a common fact of life which attracted little attention.

PRESERVED CARS

AFTER Glasgow Corporation bade farewell to its trams in September 1962, the familiar street track and stone setts gave way to slick tarmac surfaces and the cars themselves were broken up for scrap at Coplawhill Works, where most had been designed and constructed. This was a sad end indeed for a fleet of trams which had given their owners and passengers faithful service over a very long period.

However, even in 1962, when winds of change and modernisation were sweeping the country, there were those who appreciated that it would be worthwhile, even essential, to preserve some of the city's trams for the enjoyment and education of future generations. Thus, we are fortunate that in the micro-chip years of the late 20th century, no fewer than 18 preserved Glasgow trams survive, either in this country or overseas. Thanks especially to the efforts of the Scottish Tramway & Transport Society and the National Tramway Museum at Crich, we can still experience the unique sensation of travelling aboard 'Coronation' or 'Standard' tramcars from the streets of Glasgow. Furthermore, in 1988, as reported elsewhere in these pages, Glasgow trams will once again be running in their home city, for the first time in nearly 30 years.

All the cars pictured in this chapter have a single factor in common -- they are all still with us today and may be admired in the care of Museums and dedicated preservationists. But for their efforts, pictures like these would be our only 'reminders' of the Corporation's tramway system. Glasgow's trams are particularly well represented in preservation, and this is due in no small measure to the fact that the system lasted until 1962, which in UK tramway terms was very late indeed. Glasgow's was the last operational street tramway in the United Kingdom. excepting, of course, the Blackpool and Fleetwood system, which is still at work today, and highly popular at that with tourists

Above: This was an emotional occasion - the final Grand procession of cars staged on Tuesday September 4 1962. The first electric car had run in Glasgow's streets in 1898; now it was all over. Even so, many older residents still speak with affection about the cars and it seems their memory will never be allowed to die. 'Standard' car 779, in the charge of an immaculately attired motorman, is seen near Bridgeton Cross. *(FP100S)*

and local residents alike.

The last street tramway to survive 'south of the border' could be found in Sheffield, whose fine fleet of four-wheeled cars ran for the last time in 1960. Other major systems had disappeared much earlier; London (the biggest system in the country) lost its trams in 1952, Birmingham followed its example in 1953, Liverpool in 1957 and Leeds in 1959. In Scotland, Edinburgh and Dundee abandoned tramways in 1956 and Aberdeen followed in 1958. Thus, when Glasgow's trams finally ceased running in 1962, it was an emotionally-charged occasion. The first electric trams had entered service between Mitchell Street and Springburn in October 1898, in drenching rain on September 4 1962, the story was brought to an end It has been estimated that around 250,000 people thronged the city streets in the rain to pay their own last

respects to the Glasgow tram. The event was widely reported in the press and on September 5, after the last car had been stabled in its depot for the last time, the cars must have been oddly conspicuous by their absence, with the rails abandoned and the overhead wires standing in silent testimony to a bygone era. One day the trams were there, as it appeared that they had always been, the next day, they had gone.

Fortunately, Glasgow had decided to establish its own Transport Museum, including not only buses, but also trams, cars and even steam locomotives. The city boasted a proud tradition in public transport, heavy engineering and steam locomotive construction, and the new Museum was conceived to salute and perpetuate this tradition. Thus, the former paintshop at the Corporation's Coplawhill tram workshops was chosen, and the Museum was formally opened by HM Queen Elizabeth The Queen Mother, on April 14 1964. Displayed inside were seven of the City's trams.

The preserved fleet of Glasgow tramcars is housed as follows:

GLASGOW MUSEUM OF TRANSPORT:

HORSE TRAM 543: Built 1896 and preserved in substantially original condition.

SINGLE DECK ('ROOM & KITCHEN') CAR 672: Built 1898 and preserved in substantially original condition.

'STANDARD' CAR 779: Built 1900 and preserved in 1910 condition, with open balcony on top deck and open driving platform.

'STANDARD' CAR 1088: Built in 1924 and preserved today in 1930s condition, with 'blue' service indication.

EXPERIMENTAL SINGLE DECK BOGIE CAR 1089: Built in 1926 and preserved in substantially original condition.

'CORONATION' CAR 1173: Built in 1938 and preserved in substantially original condition.

MK II 'CORONATION' (or 'CUNARDER') CAR 1392: Introduced in 1952 and preserved in 1957 condition.

NATIONAL TRAMWAY MUSEUM, CRICH, DERBYSHIRE:

'STANDARD' CAR 22: Built in 1922, and although subsequently modified, it was rebuilt prior to withdrawal in original condition, with open balconies on the top deck. This tram was transported to Glasgow in 1988 for use on the Garden Festival tramway.

'STANDARD' CAR 812: Built in 1900, but appears today in 1930 condition, with a 'yellow' service indication.

EX-PAISLEY CAR 68; (Later No. 1068 in Glasgow fleet): Built in 1919 and preserved in near-original condition. Transported to Glasgow in 1988 for use at the Garden festival.

EXPERIMENTAL BOGIE CAR 1100: Built in 1928, this is a modified 'Kilmarnock Bogie' car, rebuilt 1940/41 to resemble a 'Coronation' car.

'STANDARD' DOUBLE BOGIE CAR 1115: Built in 1929, and preserved in 1930s condition with a 'red' service indication.

'CORONATION' CAR 1282: Built in 1940, and preserved in 1952/53 condition.

MK II 'CORONATION' (or 'CUNARDER') 1297: Built in 1948 and preserved in original condition. Moved to Glasgow in 1988 for use at the Garden festival.

PARIS TRANSPORT MUSEUM:

'STANDARD' CAR 488: Built in 1903 and preserved in 1950 condition.

THE SCIENCE MUSEUM, SOUTH KENSINGTON, LONDON.

'STANDARD' CAR 585: Built in 1901 and preserved in 1930 condition, with 'blue' route indication.

SEASHORE TROLLEY MUSEUM, KENNEBUNKPORT, MAINE, USA.

'CORONATION' CAR 1274: Built in 1940 and preserved in post-1957 condition.

EAST ANGLIA TRANSPORT MUSEUM, LOWESTOFT, SUFFOLK.

'CORONATION' CAR 1245: Built in 1939 and preserved in post-1957 condition.

The pictures accompanying this chapter illustrate some of these preserved Glasgow trams in their days as working cars for the Corporation. 'Standard' car No. 22, preserved today in its early working condition, is also illustrated on the rear cover of this book.

Left: In April 1959, 'Standard' car No. 812 is seen in Pollokshaws Road, at the junction of Bengal Street. The driver of TGE 687 was clearly in a hurry and taking rather a risk by overtaking two moving vehicles -it is to be hoped that he didn't meet another tram travelling in the opposite direction! The photographer has allowed the tram to clear the splendid advertisement for *Senior Service* cigarettes before pressing the shutter, and whilst this may not have been intentional, it has improved the picture considerably. Further down the street are advertisements for Ushers Ales and the Barrowland Market.
(FP101S)

Right: As a consequence of a cruel twist of fate, 'Standard' car No. 488, shown here at Baillieston in April 1961, became one of the very last 'Standard' cars to run in public service. This tram had already been withdrawn and repainted in readiness for its transfer to the Paris Transport Museum. However, as a result of the disastrous fire at Dalmarnock Depot on March 22 1961, in which 50 cars were destroyed, No. 488 was therefore returned to service, albeit temporarily, with sister 'Standards' Nos. 76, 585, 1051 and 1088. Sadly, the fire at Dalmarnock also deprived posterity of several prizes, for the flames consumed 'Standards' 108 and 526, both listed for preservation. *(FP102)*

Above: A pedestrian is clearly in a hurry in April 1960, sprinting across the setts in front of 'Coronation' car No. 1274, approaching the junction of Hayburn Street and Dumbarton Road. Standing on the right is the uniformed pointsman from Partick Depot. Three aspects of this view at least are still with us today: Kellogg's Corn Flakes (as advertised on the hoarding in the upper left corner of the view), the seemingly timeless Leyland 'Mini' saloon and singer Cliff Richard, who according to the advertisement on the railway bridge, was starring in *Expresso Bongo* at the Rosevale! No. 1274 is preserved in the United States of America.

Below:Coplawhill Works in the Autumn of 1962, and 'Standard' car No. 22 (built in 1922) is being rebuilt with open upper balconies, in the style in which it ran as a 'white' service car in the 1920s. 'Standard' No. 22 was rebuilt thus at Coplawhill following the final procession of September 4 1962, in readiness for its subsequent preservation. The car is now owned by the National Tramway Museum but in 1988 it returned to Glasgow to work on the Garden Festival Tramway, as illustrated on the back cover of this book. To the right of No. 22 in this view is 'Standard' 585, which is undergoing a rather less extensive restoration as a 'blue' service car in the style of the 1930s. This car also escaped the scrapman's torch and today car No. 585 can be seen amidst a fascinating array of technology of many varieties in the Science Museum, South Kensington. *(FP104)*

Left: Another scene inside Coplawhill Works in September 1962, with experimental single deck car No. 1089 and 'Coronation' No. 1173 in immaculate condition, awaiting their final call to service in the procession of September 4. No.1089 was designed in 1925 on a 'passenger flow' system working on the principle that the car would have a rear entrance and front exit. However, as with other experiments of this nature, it was not successful and the 'one way' system was abandoned in 1932. The car incorporated many modern features, including air brakes, and could travel at relatively high speed. services on which No. 1089 was used in its early career included Langside-Sinclair Drive, Johnstone-Paisley Cross and the Clydebank-Duntocher shuttle. After the closure of this last-named service in 1949, No. 1089 languished in Langside Depot and then Coplawhill Works for the next two years, prior to re-entering service with half of its longitudinal seats removed. This modification enabled the car to provide maximum capacity for peak hour shipyard services.No. 1089 could accommodate 20 seated and 38 standing passengers and was withdrawn in June 1961; it was stored at Partick for a year prior to the final procession. *(FP105)*

Left: This picture of 'Cunarder' No. 1297, seen in Argyle Street in April 1960, encapsulates the spirit of Glasgow's trams in their latter years. Luckily, preservation has ensured that some cars are more than just a treasured memory. *(FP106)*

Below left: Here, 25 years later and some 200 miles away from its former home, No. 1297 poses in a very well known location! The car was loaned by the National Tramway Museum at Crich in Derbyshire in 1984 for a two-year period to celebrate the Blackpool tramway centenary in 1985.

With its 4 x 36 hp motors No. 1297 could chase seagulls on the more open stretches of track! It was very popular with the Blackpool tram crews, but on occasions put the wind up them with its seeming reluctance to stop! One of 100 similar trams built between 1948 and 1952, together with sister car No. 1392 it survived breaking up. No. 1297 returned to the Tramway Museum early in 1986, and No. 1392 now resides in the Museum of Transport at the Kelvin Hall in Glasgow. *(A. Stevenson)*

Above right: With a splendidly casual indifference to modern traffic flow, Hex-Dash 'Standard' car No. 1088, the last 'Standard' to be built (in 1926) rumbles along Cumbernauld Road, passing Dee Street, on August 10 1955. Service 6 at this time ran from Alexandra Park or Riddrie to Scotstoun, and survived until November 1 1959, when it was replaced by motorbus service 56. No. 1088 also now resides in the Museum of Transport at the Kelvin Hall. *(R. J. S. Wiseman)*

Right: Another photograph of experimental single-decker No. 1089, this time at Dalmuir West on a shipyard special one late Friday afternoon in April 1953. Built at Coplawhill in 1926, its specification called for 'a high-speed pullman single deck bogie car with passenger front exit and rear entrance', and it was completed in that form. Following various internal and external body rearrangements over the years, it made its final journey in public service in mid-1961, but was to appear in the closing farewell procession on September 4 1962, and now also resides in Glasgow's Museum of Transport.

Coincidentally, 'Kilmarnock bogie' car No. 1115 behind it, waiting its turn to use the crossover to make the cross-city journey back to Auchenshuggle on service 9, also survived, and is preserved at Crich. *(R. J. S. Wiseman)*

'Coronation' Mark 1 car No. 1282 shimmers in the midday sun at Renfrew Ferry on Wednesday July 27 1955. Service 28 was synonymous with the Paisley system, which was finally abandoned on May 11 1957. Always a heavily trafficked route, it earned the nickname of 'The Goldmine'.

The advert on the car's side panel shows that David Blane & Son Ltd or Weir Street Garage, Paisley, was the agent for Jowett Javelin cars (years ahead of their time) and Bradford vans, as well as Jaguar, Standard and Triumph cars.

No. 1282 had a special claim to fame. Not to be outdone by Glasgow, which held its final tram procession on September 4 1962, Clydebank went one better using No. 1282 to run from the Town Hall to Dalmuir West, Yoker and back two days later, thus making it then the last tram to run in Scotland. Happily it survived, and was subsequently moved to the Tramway Museum at Crich, where she remains to this day. (R. J. S. Wiseman)

BIBLIOGRAPHY

The Glasgow Tramcar, Ian Stewart

The Glasgow Coronations, Light Railway Transport League

Glasgow's Trams, The Glasgow Museum of Transport

The National Tramway Museum, Tramway Museum Society, Crich, Derbyshire

A Handbook of Glasgow Trams, D. L. Thomson

The Howden Quarterly, J. F. Howden & Co. Ltd., Glasgow

The Golden Age of Tramways, Charles Klapper

The Green Goddesses Go East, Ian L. Cormack

Periodicals

Scottish Transport, various issues

The Scottish Daily Express, various issues

'. . .With so many fond memories retained, it is safe to say that many a Glasgow citizen, if he could be assured that when his time comes he could stand at the edge of a straight stretch of celestial road and see a Glasgow tramcar of his favourite colour looming in the distance along the shining lines, he would be more than supremely happy. He would know he had arrived in Heaven.'

Reproduced from *The Howden Quarterly*, Winter 1962-63, courtesy of James Howden & Co. Ltd., Glasgow